Holt Mathematics

Chapter 11 Resource Book

HOLT, RINEHART AND WINSTON
A Harcourt Education Company
Orlando • Austin • New York • San Diego • London

Copyright © by Holt, Rinehart and Winston

All rights reserved. No part of this publication may be reproduced or transmitted in any form or by any means, electronic or mechanical, including photocopy, recording, or any information storage and retrieval system, without permission in writing from the publisher.

Teachers using HOLT MATHEMATICS may photocopy complete pages in sufficient quantities for classroom use only and not for resale.

Printed in the United States of America

If you have received these materials as examination copies free of charge, Holt, Rinehart and Winston retains title to the materials and they may not be resold. Resale of examination copies is strictly prohibited and is illegal.

Possession of this publication in print format does not entitle users to convert this publication, or any portion of it, into electronic format.

ISBN 0-03-078402-6

6 7 8 170 09 08

CONTENTS

Blackline Masters

Parent Letter	1
Lesson 11-1 Practice A, B, C	3
Lesson 11-1 Reteach	6
Lesson 11-1 Challenge	7
Lesson 11-1 Problem Solving	8
Lesson 11-1 Reading Strategies	9
Lesson 11-1 Puzzles, Twisters & Teasers	10
Lesson 11-2 Practice A, B, C	11
Lesson 11-2 Reteach	14
Lesson 11-2 Challenge	16
Lesson 11-2 Problem Solving	17
Lesson 11-2 Reading Strategies	18
Lesson 11-2 Puzzles, Twisters & Teasers	19
Lesson 11-3 Practice A, B, C	20
Lesson 11-3 Reteach	23
Lesson 11-3 Challenge	25
Lesson 11-3 Problem Solving	26
Lesson 11-3 Reading Strategies	27
Lesson 11-3 Puzzles, Twisters & Teasers	28
Lesson 11-4 Practice A, B, C	29
Lesson 11-4 Reteach	32
Lesson 11-4 Challenge	33
Lesson 11-4 Problem Solving	34
Lesson 11-4 Reading Strategies	35
Lesson 11-4 Puzzles, Twisters, & Teasers	36
Lesson 11-5 Practice A, B, C	37
Lesson 11-5 Reteach	40
Lesson 11-5 Challenge	42
Lesson 11-5 Problem Solving	43
Lesson 11-5 Reading Strategies	44
Lesson 11-5 Puzzles, Twisters & Teasers	45
Lesson 11-6 Practice A, B, C	46
Lesson 11-6 Reteach	49
Lesson 11-6 Challenge	51
Lesson 11-6 Problem Solving	52
Lesson 11-6 Reading Strategies	53
Lesson 11-6 Puzzles, Twisters & Teasers	54
Answers to Blackline Masters	84

Date _____

Dear Family,

In this chapter, your child will learn more about how to solve equations and inequalities. These skills are needed in science, in business, and in everyday life.

A **multi-step equation** is an equation that requires more than one operation to solve.

Carly had a $10 gift certificate for her favorite restaurant. After she paid the bill plus a 20% tip, the $10 was deducted. The amount she paid was $4.40. What was her original bill?

Let b represent the amount of the original bill.

$b + 0.20b - 10 = 4.40$ *bill + tax − gift certificate = amount paid*
$1.20b - 10 = 4.40$ *Combine like terms.*
$\underline{ + 10 +10}$ *Add 10 to both sides.*
$1.20b = 14.40$
$\dfrac{1.20b}{1.20} = \dfrac{14.40}{1.20}$ *Divide both sides by 1.20.*
$b = 12$ *Her original bill was $12.*

Solving a **multi-step inequality** uses the same operations as solving a multi-step equation. However, if both sides of the inequality are multiplied or divided by a negative number, the inequality symbol must be reversed.

The school band plans to march in a parade in another town. They raised $650 at a carwash, but they estimate that the entire cost of the trip will be $1400. If they sell calendars for $7.50 each, how many calendars must they sell to make enough money for their trip?

The money made from selling x calendars at $7.50 each is $7.50x$. The money the band will have is the money made from selling calendars plus the money raised at the carwash, or $7.50x + 650$. This total must be at least 1400.

$7.50x + 650 \geq 1400$ Let x represent the number of calendars sold.
$ - 650 - 650$ Subtract 650 from both sides.
$7.50x \geq 750$
$\dfrac{7.50x}{7.50} \geq \dfrac{750}{7.50}$ Divide both sides by 7.50.
$x \geq 100$

The band must sell at least 100 calendars to make enough money for their trip.

Your child will also learn how to solve **systems of equations**. A system of equations is a set of two or more equations with a common solution.

$$y = 2x - 10$$
$$y = 2 - x$$

Since this system has two variables, x and y, the solution will have two values that can be written as an **ordered pair**, (x, y).

Because the expressions $2x - 10$ and $2 - x$ both equal y, they can be used to write an equation in one variable.

$y = 2x - 10 \qquad\qquad y = 2 - x$

$$
\begin{aligned}
2x - 10 &= 2 - x \\
+ x & \quad\quad + x \qquad \text{Add x to both sides.} \\
3x - 10 &= 2 \\
+ 10 & \quad\quad + 10 \qquad \text{Add 10 to both sides.} \\
3x &= 12 \\
\frac{3x}{3} &= \frac{12}{3} \qquad \text{Divide both sides by 3.} \\
x &= 4
\end{aligned}
$$

To find y, substitute the solution for x into one of the original equations.

$$y = 2 - x \quad \rightarrow \quad y = 2 - 4$$
$$y = -2$$

Write the solution as an ordered pair: $(4, -2)$.

Your child can check the solution by substituting the $x-$ and $y-$values into the original equations.

$$
\begin{array}{ll}
y = 2x - 10 & y = 2 - x \\
-2 = 2(4) - 10 & -2 = 2 - 4 \\
-2 = -2 \checkmark & -2 = -2 \checkmark
\end{array}
$$

For additional resources, visit go.hrw.com and enter the keyword MT7 Parent.

Name _____ Date _____ Class _____

LESSON 11-1 Practice A
Simplifying Algebraic Expressions

Combine like terms.

1. $6s - 4s$

2. $3k + 3k$

3. $9b - 5b$

4. $10x - 3x$

5. $9b + 12b$

6. $8m + 3 - m$

7. $11d - 6d + 4$

8. $7r + 9r - 3$

9. $15q - 8q - 6p$

10. $3y + x - 2y$

11. $9h - 3h + 6g$

12. $7a - 4a + 2b + 3b$

Simplify.

13. $2(x + 3) + x$

14. $5(1 + y) - 3y$

15. $3(a + 3) + 1$

Solve.

16. $3a + a = 16$

17. $9x - 3x = 30$

18. $5w + 3w = 24$

19. Last Saturday Nakesha rented 4 movies from the video store. This Saturday she rented 2 movies. Let x represent the cost of renting each movie. Write and simplify an expression for how much more Nakesha spent last week renting movies.

20. If it costs $4 to rent each movie, how much more money did Nakesha spend last week renting movies?

Name _____ Date _____ Class _____

LESSON 11-1
Practice B
Simplifying Algebraic Expressions

Combine like terms.

1. $8a - 5a$

2. $12g + 7g$

3. $4a + 7a + 6$

4. $6x + 3y + 5x$

5. $10k - 3k + 5h$

6. $3p - 7q + 14p$

7. $3k + 7k + 5k$

8. $5c + 12d - 6$

9. $13 + 4b + 6b - 5$

10. $4f + 6 + 7f - 2$

11. $x + y + 3x + 7y$

12. $9n + 13 - 8n - 6$

Simplify.

13. $4(x + 3) - 5$

14. $6(7 + x) + 5x$

15. $3(5 + 3x) - 4x$

Solve.

16. $6y + 2y = 16$

17. $14b - 9b = 35$

18. $3q + 9q = 48$

19. Gregg has q quarters and p pennies. His brother has 4 times as many quarters and 8 times as many pennies as Gregg has. Write the sum of the number of coins they have, and then combine like terms.

20. If Gregg has 6 quarters and 15 pennies, how many total coins do Gregg and his brother have?

Name _____ Date _____ Class _____

LESSON 11-1 Practice C
Simplifying Algebraic Expressions

Combine like terms.

1. $7x + 3x + 5x$

2. $9a + 4a - 8a$

3. $12g + 10h - 3g + h$

4. $5b + 7 + 7b - 6$

5. $3p + 6q - p - 3q$

6. $11s - 6s + 9 + 3s$

7. $9 + 3f + 8f - 6 - 3f$

8. $v + 4y - 2y + 5y$

9. $15k + 8m - 7k + 2m$

10. $2 + 6a + 9b - 3b - a$

11. $11j + 8m - 6n + 3$

12. $13x + 9y - 6x - 8y + 1$

Simplify.

13. $5(y + 6) + 4$

14. $6(3x - 2) + 4x$

15. $3(2 + x) + 6x - 4$

Solve.

16. $3x + 8 + 5x = 48$

17. $4(g + 7) = 64$

18. $9a - 6 = 3a + 4 + 4a$

19. Suppose the lengths of the four sides of a quadrilateral are represented by the expressions $3a$, $a + 2$, $2a - 1$, and $3a + 6$. Write the sum of the lengths (the perimeter), and then simplify.

20. Find the perimeter of the quadrilateral in Exercise 19 for $a = 4$ in.

LESSON 11-1 Reteach
Simplifying Algebraic Expressions

The parts of an expression separated by plus or minus signs are called **terms**.

The expression shown has four terms. You can combine two of these terms to **simplify** the expression.

$$5a + 7b - 3a + 6a^2$$
$$5a - 3a + 7b + 6a^2$$
$$2a + 7b + 6a^2$$

$$5a + 7b - 3a + 6a^2$$

Like terms have the same variable raised to the same power.

Equivalent expressions have the same value for all values of the variables.

Some algebraic equations can be solved by first combining like terms.
Solve $4w - w = 24$.

$$4w - w = 24$$ — Identify like terms; w is $1w$.
$$3w = 24$$ — Combine coefficients of like terms.
$$\frac{3w}{3} = \frac{24}{3}$$ — Divide both sides by 3.
$$w = 8$$

Complete to combine like terms.

1. $9z + 4z$

   $(9 + \underline{})z$

   $\underline{}z$

2. $9r + 5q - 2r$

   $(9 - \underline{})r + 5q$

   $\underline{}r + 5q$

3. $5t + 12f - t - 3f$

   $(5 - \underline{})t + (12 - \underline{})f$

   $\underline{}t + \underline{}f$

Simplify.

4. $7m + 3n - m + 2n$

5. $15r + 4 - 3r - 2$

6. $6x + 3z - y$

Complete to solve.

7. $5h + 2h = 21$

   $\underline{}h = 21$

   $\dfrac{\underline{}h}{\underline{}} = \dfrac{21}{\underline{}}$

   $h = \underline{}$

8. $16w - 5w = 44$

   $\underline{}w = 44$

   $\dfrac{\underline{}w}{\underline{}} = \dfrac{44}{\underline{}}$

   $w = \underline{}$

9. $48 = 13x - x$

   $48 = \underline{}x$

   $\dfrac{48}{\underline{}} = \dfrac{\underline{}x}{\underline{}}$

   $\underline{} = x$

Holt Mathematics

Challenge

11-1 Mission Operation

You can create your own operation. Take a symbol, such as ▲, and tell what that symbol means using the standard operations of +, −, ×, and ÷.

Example
If $x \triangle y = x + 2xy$, find $5 \triangle 3$.

$x \triangle y = x + 2xy$ Use the operation as defined.
$5 \triangle 3 = 5 + 2(5)(3)$ Substitute 5 for x and 3 for y.
$ = 5 + 30$ Carry out the standard operations.
$ = 35$

Apply the definition of ▲. First, show how to substitute the values or expressions that replace x and y. Then, carry out the standard operations to simplify completely.

1. $7 \triangle 2$

2. $2 \triangle 7$

3. $(a \triangle b) + (b \triangle a)$

4. $[a \triangle (4b)] + [(2a) \triangle b]$

Let $a ✳ b = 2a + \dfrac{b}{2}$. Apply the definition of ✳. Simplify the resulting expression.

5. $(9 ✳ 4) + (3 ✳ 8)$

6. $(r ✳ 4t) + (4r ✳ 2t)$

Problem Solving
11-1 Simplifying Algebraic Expressions

Write the correct answer.

1. An item costs x dollars. The tax rate is 5% of the cost of the item, or $0.05x$. Write and simplify an expression to find the total cost of the item with tax.

2. A sweater costs d dollars at regular price. The sweater is reduced by 20%, or $0.2d$. Write and simplify an expression to find the cost of the sweater before tax.

3. Consecutive integers are integers that differ by one. You can represent consecutive integers as x, $x + 1$, $x + 2$ and so on. Write an equation and solve to find three consecutive integers whose sum is 33.

4. Consecutive even integers can be represented by x, $x + 2$, $x + 4$ and so on. Write an equation and solve to find three consecutive even integers whose sum is 54.

Choose the letter for the best answer.

5. In Super Bowl XXXV, the total number of points scored was 41. The winning team outscored the losing team by 27 points. What was the final score of the game?
    A 33 to 8
    B 34 to 7
    C 22 to 2
    D 18 to 6

6. A high school basketball court is 34 feet longer than it is wide. If the perimeter of the court is 268, what are the dimensions of the court?
    F 234 ft by 34 ft
    G 67 ft by 67 ft
    H 70 ft by 36 ft
    J 84 ft by 50 ft

7. Julia ordered 2 hamburgers and Steven ordered 3 hamburgers. If their total bill before tax was $7.50, how much did each hamburger cost?
    A $1.50
    B $1.25
    C $1.15
    D $1.02

8. On three tests, a student scored a total of 258 points. If the student improved his performance on each test by 5 points, what was the score on each test?
    F 81, 86, 91
    G 80, 85, 90
    H 75, 80, 85
    J 70, 75, 80

Reading Strategies
11-1 Organization Patterns

A statement written with numbers and words, such as

    3 apples + 2 pears + 4 bananas + 2 apples + 6 bananas

can be rewritten with numbers and variables, like

    $3a + 2p + 4b + 2a + 6b$.

Like variables can then be combined.

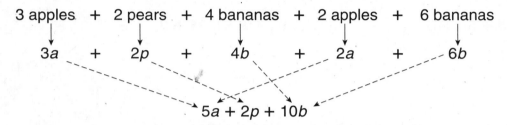

In an expression, **terms** are separated by + and − signs.
The expression below has 5 terms:

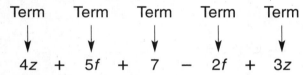

Reorganize the terms so that $(4z + 3z) + (5f − 2f) + 7$
like terms are together.
Combine like terms. $7z + \quad 3f + 7$

Answer each question.

1. Rewrite this statement with numbers and variables:
   3 kites + 4 bats + 2 kites + 3 bats.

   _____

2. How many terms are in the statement above?

   _____

3. Reorganize terms so like terms are next to each other.

   _____

4. Combine like terms.

   _____

Name _____ Date _____ Class _____

LESSON 11-1 Puzzles, Twisters & Teasers
Buried Treasure

Combine like terms.
Shade in the answers on the grid to reveal a hidden picture.
Hint: The picture is a drawing on a map made by a famous pirate.

1. $6x - 2x$

2. $14p - 8 - 5p$

3. $4n + 5n + 7$

4. $2h + 3n + 3h + 5n + 6$

5. $9x + 8y + 2x + 3y - 8$

Simplify.
Shade the answers on the grid to continue to reveal the picture.

6. $8y - 7y$ _____

7. $4x + 7 - 3x$ _____

8. $13y - 2y + 2$ _____

9. $5x + 3x + 4$ _____

10. $2x - 6 + 3x$ _____

11. $12y - 4y$ _____

12. $5x + 2x + 3$ _____

13. $4d + 2d - 3$ _____

$9p - 8$	$10x$	$5z + 17$	$8f + 2g - 14$	$12y - 16$	$7g + 5h - 12$	$9n + 7$
$6y - 5t + 16$	y	$4a + 8$	$3x - 2$	$5p$	$x + 7$	$17g$
$17h - 8g + 16$	$7z + 8b - 7$	$11y + 2$	$6y + 8$	$8y$	$5a + 2x$	$7x - 9$
$5x - 3$	$19x - 4y + 8$	$18y + 3$	$6d - 3$	$14c + 18$	$7x + 6y - 8$	$4a + 8$
$10x - 2$	$15v + 6y + 3$	$8x + 4$	$9h - 7n + 5$	$5h + 8n + 6$	$12x - 4$	$34b + 5$
$3d - 6h + 14$	$11x + 11y - 8$	$3c + 4n - 2$	$6d + 4e - 12$	$9a + 6v - 3$	$5x - 6$	$12c + 13$
$4x$	$9n + 5z - 23$	$8x - 3$	$5x + 48$	$16y - 4n + 5$	$7t + 56 - 3g$	$7x + 3$

Practice A
LESSON 11-2 Solving Multi-Step Equations

Describe the operations used to solve the equation.

1. $4x + 6 - 2x = 14$

   $2x + 6 = 14$ _____

   $2x + 6 - 6 = 14 - 6$ _____

   $2x = 8$

   $\dfrac{2x}{2} = \dfrac{8}{2}$ _____

   $x = 4$

2. $\dfrac{5x}{6} + \dfrac{2x}{3} = 3$

   $6\left(\dfrac{5x}{6} + \dfrac{2x}{3}\right) = 6(3)$ _____

   $5x + 4x = 18$

   $9x = 18$ _____

   $\dfrac{9x}{9} = \dfrac{18}{9}$ _____

   $x = 2$

Solve.

3. $2x - 8 + 3x = 7$

4. $6y + 9 - 4y = -3$

5. $5a - 1 + a = 11$

_____ _____ _____

6. $3w - 3 + 2w = -18$

7. $8 - 2d + 4d = 12$

8. $-3 = 4x + 6 - x$

_____ _____ _____

9. $\dfrac{1}{4} = \dfrac{r}{4} - \dfrac{7}{4} + \dfrac{3r}{4}$

10. $\dfrac{2x}{3} + \dfrac{1}{3} - \dfrac{x}{3} = \dfrac{2}{3}$

11. $\dfrac{3n}{5} + \dfrac{9}{10} + \dfrac{n}{5} = -2\dfrac{3}{10}$

_____ _____ _____

12. Together Marc and Paoli have $32. Marc has $4 less than twice the amount Paoli has. How much does each man have?

_____

Holt Mathematics

LESSON 11-2 Practice B
Solving Multi-Step Equations

Solve.

1. $2x + 5x + 4 = 25$

2. $9 + 3y - 2y = 14$

3. $16 = 4w + 2w - 2$

4. $26 = 3b - 2 - 7b$

5. $31 + 4t - t = 40$

6. $14 - 2x + 4x = 20$

7. $\frac{5m}{8} - \frac{6}{8} + \frac{3m}{8} = \frac{2}{8}$

8. $-4\frac{2}{3} = \frac{2n}{3} + \frac{1}{3} + \frac{n}{3}$

9. $7a + 16 - 3a = -4$

10. $\frac{x}{2} + 1 + \frac{3x}{4} = -9$

11. $7m + 3 - 4m = -9$

12. $\frac{2x}{5} + 3 - \frac{4x}{5} = \frac{1}{5}$

13. $\frac{7k}{8} - \frac{3}{4} - \frac{5k}{16} = \frac{3}{8}$

14. $6y + 9 - 4y = -3$

15. $\frac{5a}{6} - \frac{7}{12} + \frac{3a}{4} = -2\frac{1}{6}$

16. The measure of an angle is 28° greater than its complement. Find the measure of each angle.

17. The measure of an angle is 21° more than twice its supplement. Find the measure of each angle.

18. The perimeter of the triangle is 126 units. Find the measure of each side.

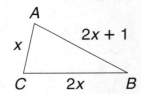

19. The base angles of an isosceles triangle are congruent. If the measure of each of the base angles is twice the measure of the third angle, find the measure of all three angles.

Practice C
Lesson 11-2: Solving Multi-Step Equations

Solve.

1. $54 + 8x + 7x = 21$

2. $\frac{2y}{3} + 27 + \frac{3y}{4} = -24$

3. $18x - 21 - 15x = 3$

4. $1.15a + 8 - 0.4a = -7$

5. $3 = 13w - 9w - 8$

6. $\frac{4m}{9} + 13 - \frac{1m}{3} = 4$

7. $182d + 3.5 - 20d = 19.7$

8. $\frac{3k}{4} - 28 - \frac{2k}{3} = 37$

9. $\frac{3x}{4} - \frac{2x}{3} = \frac{2}{5}$

10. Takeo and Manuel drove to a baseball game together. The tickets cost $10.50 each and parking was $6. Popcorn cost $2.25 a box and each of the boys bought a box of popcorn. Takeo bought 1 drink and Manuel bought 2 drinks. The baseball game and the snacks cost the boys a total of $39.75. How much did each of the drinks cost?

11. The perimeter of right triangle ABC is 132 units. Find the length of the hypotenuse.

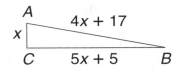

12. Mrs. Lincoln budgets her monthly take-home pay. The remainder she deposits in her savings account. She budgets $\frac{1}{3}$ of her take-home pay for rent, $\frac{1}{8}$ for car expenses, $\frac{1}{6}$ for food, $\frac{1}{24}$ for insurance and $\frac{1}{16}$ for entertainment. What is Mrs. Lincoln's monthly take-home pay if she deposits $546 in her savings account?

Reteach

11-2 Solving Multi-Step Equations

To combine like terms, add (or subtract) coefficients.

$2m + 3m = (2 + 3)m = 5m$ $x - 3x = (1 - 3)x = -2x$

To solve an equation that contains like terms, first combine the like terms.

$2m + 3m = 35 - 25$ **Check:** Substitute into the original.

$5m = 10$ Combine like terms. $2m + 3m = 35 - 25$

$\dfrac{5m}{5} = \dfrac{10}{5}$ Divide by 5. $2(2) + 3(2) \stackrel{?}{=} 35 - 25$

$m = 2$ $4 + 6 \stackrel{?}{=} 10$

 $10 = 10$ ✓

$x + 6 - 3x + 5 = 13$ **Check:** $x + 6 - 3x + 5 = 13$

$-2x + 11 = 13$ Combine like terms. $-1 + 6 - 3(-1) + 5 \stackrel{?}{=} 13$

$\underline{-11 -11}$ Subtract 11. $-1 + 6 + 3 + 5 \stackrel{?}{=} 13$

$-2x = 2$ $-1 + 14 \stackrel{?}{=} 13$

$\dfrac{-2x}{-2} = \dfrac{2}{-2}$ Divide by -2. $13 = 13$ ✓

$x = -1$

Complete to solve and check each equation.

1. $4z - 7z = -20 - 1$ **Check:** $4z - 7z = -20 - 1$

   $\underline{}z = \underline{}$ Combine like terms. $4(\underline{}) - 7(\underline{}) \stackrel{?}{=} -20 - 1$

   $\dfrac{\underline{}z}{} = \dfrac{\underline{}}{}$ Divide. $\underline{} - \underline{} \stackrel{?}{=} \underline{}$

   $z = \underline{}$

2. $t + 1 - 4t + 8 = 21$ **Check:** $t + 1 - 4t + 8 = 21$

   $\underline{}t \underline{} = \underline{}$ Combine like terms. $\underline{} + 1 - 4(\underline{}) + 8 \stackrel{?}{=} \underline{}$

   $\underline{} \underline{}$ Subtract. $\underline{} + 1 + \underline{} + 8 \stackrel{?}{=} 21$

   $\underline{}t = \underline{}$ $\underline{} + \underline{} \stackrel{?}{=} 21$

   $\dfrac{\underline{}t}{} = \dfrac{\underline{}}{}$ Divide.

   $t = \underline{}$

Reteach
11-2 Solving Multi-Step Equations (continued)

To clear fractions in an equations, multiply every term by the least common denominator (LCD)

$$\frac{x}{4} + 3 = \frac{1}{2}$$ The LCD of 4 and 2 is 4.   **Check:** $\frac{1}{4}(-10) + 3 \stackrel{?}{=} \frac{1}{2}$

$$4 \cdot \frac{x}{4} + 4 \cdot 3 = 4 \cdot \frac{1}{2}$$ Multiply *every* term by LCD. $\frac{-10}{4} + \frac{12}{4} \stackrel{?}{=} \frac{1}{2}$

$$x + 12 = 2$$ $\frac{2}{4} \stackrel{?}{=} \frac{1}{2}$

$$\underline{-12 \quad -12}$$ Subtract 12. $\frac{1}{2} = \frac{1}{2}$ ✓

$$x = -10$$

Complete to solve and check.

3. $\frac{t}{2} + \frac{t}{6} = 2$ Determine the LCD.   **Check:** $\frac{1}{2}(\underline{}) + \frac{1}{6}(\underline{}) \stackrel{?}{=} 2$

$\underline{} \cdot \frac{t}{2} + \underline{} \cdot \frac{t}{6} = \underline{} \cdot 2$ Multiply *every* term by LCD. $\frac{}{2} + \frac{}{2} \stackrel{?}{=} 2$

$\underline{} t + t = \underline{}$ The fractions are cleared. $\frac{}{2} \stackrel{?}{=} 2$

$\underline{} t = \underline{}$ Combine like terms.

$\frac{\underline{}t}{} = \frac{\overline{}}{\underline{}}$ Divide.

$t = \underline{}$

4. $\frac{m}{3} + \frac{m}{4} - 1 = \frac{5}{2}$ Determine the LCD.

$\underline{} \cdot \frac{m}{3} + \underline{} \cdot \frac{m}{4} - \underline{} \cdot 1 = \underline{} \cdot \frac{5}{2}$ Multiply *every* term by the LCD.

$\underline{} m + \underline{} m - \underline{} = \underline{}$ The fractions are cleared.

$\underline{} m - \underline{} = \underline{}$ Combine like terms.

Add.

$\underline{} m = \underline{}$

$\frac{\underline{}m}{} = \frac{\overline{}}{\underline{}}$ Divide.

$m = \underline{}$ **Check:** $\frac{}{3} + \frac{}{4} - 1 \stackrel{?}{=} \frac{5}{2}$

$\frac{}{12} + \frac{}{12} - \frac{}{12} \stackrel{?}{=} \frac{}{12}$

Name _____ Date _____ Class _____

Challenge
LESSON 11-2 Use the Power of Algebra!

An equation may be used to solve a problem involving angle measure in a triangle.

In isosceles triangle ABC, the measure of vertex angle C is 30° more than the measure of each base angle. Find the measure of each angle of the triangle.

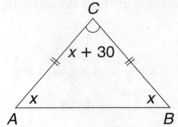

Let x = the number of degrees in m∠A.
Then x = the number of degrees in m∠B.
And $x + 30$ = the number of degrees in m∠C.

The sum of the measures of the angles of a triangle is 180°.

$x + x + x + 30 = 180$
$3x + 30 = 180$ Combine like terms.
$\underline{-30 \quad -30}$ Subtract 30.
$3x = 150$
$\dfrac{3x}{3} = \dfrac{150}{3}$ Divide by 3.
$x = 50$ ← m base ∠
$x + 30 = 80$ ← m vertex ∠

Check:
m base ∠ = 50°
m base ∠ = 50°
m vertex ∠ = $\underline{\ 80°\ }$
180° ✓

So, the measure of each base angle is 50° and the measure of the vertex angle is 80°.

Write and solve an equation to find the measures of the angles of each triangle.

1. The measure of each of the base angles of an isosceles triangle is 9° less than 4 times the measure of the vertex angle.

2. The measure of the vertex angle of an isosceles triangle is one-fourth that of a base angle.

measure of each base angle = _____
measure of vertex angle = _____

measure of each base angle = _____
measure of vertex angle = _____

Name _____ Date _____ Class _____

Problem Solving
LESSON 11-2 Solving Multi-Step Equations

A taxi company charges $2.25 for the first mile and then $0.20 per mile for each mile after the first, or $F = \$2.25 + \$0.20(m - 1)$ where F is the fare and m is the number of miles.

1. If Juan's taxi fare was $6.05, how many miles did he travel in the taxi?

   _____

2. If Juan's taxi fare was $7.65, how many miles did he travel in the taxi?

   _____

A new car loses 20% of its original value when you buy it and then 8% of its original value per year, or $D = 0.8V - 0.08Vy$ where D is the value after y years with an original value V.

3. If a vehicle that was valued at $20,000 new is now worth $9,600, how old is the car?

   _____

4. A 6-year old vehicle is worth $12,000. What was the original value of the car?

   _____

The equation used to estimate typing speed is $S = \frac{1}{5}(w - 10e)$, where S is the accurate typing speed, w is the number of words typed in 5 minutes and e is the number of errors. Choose the letter of the best answer.

5. Jane can type 55 words per minute (wpm). In 5 minutes, she types 285 words. How many errors would you expect her to make?

   A 0   C 2
   B 1   D 5

6. If Alex types 300 words in 5 minutes with 5 errors, what is his typing speed?

   F 48 wpm   H 59 wpm
   G 50 wpm   J 60 wpm

7. Johanna receives a report that says her typing speed is 65 words per minute. She knows that she made 4 errors in the 5-minute test. How many words did she type in 5 minutes?

   A 285   C 365
   B 329   D 1825

8. Cecil can type 35 words per minute. In 5 minutes, she types 255 words. How many errors would you expect her to make?

   F 2   H 6
   G 4   J 8

Copyright © by Holt, Rinehart and Winston.
All rights reserved.

Holt Mathematics

Name _____ Date _____ Class _____

LESSON 11-2 Reading Strategies
Compare and Contrast

To solve equations with many terms, you can compare terms and combine the terms that are alike on one side of the = sign.

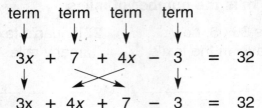

Step 1: Rearrange terms. $3x + 4x + 7 - 3 = 32$

Step 2: Combine like terms. $7x + 4 = 32$

Compare the first line of the equation with the second line.

1: How many terms are on the left of the = sign in the equation?

_____

2. How many terms are there after the terms are rearranged in Step 1?

_____

3. How are these two arrangements alike?

_____

4. How are these two arrangements different?

_____

Now compare after the like terms are combined.

5. How many terms are on the left of the = sign in Step 2?

_____

6. What is still alike and what is different about the equation now?

_____

_____

Puzzles, Twisters & Teasers
11-2 Don't Miss the Boat!

Decide whether or not the given solution to each equation is correct. Circle the letter above your answer. Then unscramble the letters to solve the riddle.

1. $2x + 4 + 5x - 8 = 24$ $x = 4$
   - **A** correct
   - **Q** incorrect

2. $\dfrac{5n}{6} - \dfrac{1}{4} = \dfrac{3}{8}$ $n = 7$
   - **P** correct
   - **D** incorrect

3. $\dfrac{3y}{7} + \dfrac{5}{7} = -\dfrac{1}{7}$ $y = -2$
   - **M** correct
   - **J** incorrect

4. $\dfrac{x}{2} + \dfrac{2}{3} = \dfrac{5}{6}$ $x = 3$
   - **Y** correct
   - **I** incorrect

5. $5y - 2 - 8y = 31$ $y = 20$
   - **Z** correct
   - **R** incorrect

6. $\dfrac{2p}{3} + \dfrac{p}{4} - \dfrac{1}{6} = \dfrac{7}{2}$ $p = 4$
   - **A** correct
   - **X** incorrect

7. $\dfrac{b}{6} + \dfrac{3b}{8} = \dfrac{5}{12}$ $b = 9$
   - **T** correct
   - **L** incorrect

8. $2a + 7 + 3a = 32$ $a = 5$
   - **B** correct
   - **L** incorrect

9. $\dfrac{h}{6} + \dfrac{h}{8} = 1\dfrac{1}{6}$ $h = 2$
   - **W** correct
   - **Y** incorrect

10. $b + \dfrac{b}{5} - 10 = 4\dfrac{2}{5}$ $b = 12$
    - **R** correct
    - **B** incorrect

Who's the head of the seagull navy?

__ __ __ __ __ __ __ __ __ __ D

Name _____ Date _____ Class _____

LESSON 11-3 Practice A
Solving Equations with Variables on Both Sides

Tell which term you would add or subtract on both sides side of the equation so that the variable is only on one side.

1. $7x - 1 = 2x + 5$

2. $3y + 1 = 4y - 6$

3. $10 - y = 2y + 3$

4. $6x + 2 = 1 - x$

Solve.

5. $3a - 10 = 4 + a$

6. $5y + 4 = 4y + 5$

7. $b + 3 = 2b - 7$

8. $2x - 1 = x - 3$

9. $2w + 3 = 4w - 5$

10. $6n + 4 = n - 11$

11. $\dfrac{19m}{5} - \dfrac{3}{5} = 1.4 - \dfrac{m}{5}$

12. $11 - \dfrac{t}{2} = \dfrac{t}{3} - 14$

13. $29r + 4 = 28 + 13r$

14. The difference of four times a number and five is the same as three more than twice a number. Find the number. _____

15. Fifteen more than twice the hours Carla worked last week is the same as three times the hours she worked this week decreased by 15. She worked the same number of hours each week. How many hours did she work each week? _____

Name _____ Date _____ Class _____

Practice B
LESSON 11-3 Solving Equations with Variables on Both Sides

Solve.

1. $7x - 11 = -19 + 3x$

2. $11a + 9 = 4a + 30$

3. $4t + 14 = \frac{6t}{5} + 7$

_____ _____ _____

4. $19c + 31 = 26c - 74$

5. $\frac{3y}{8} - 9 = 13 + \frac{y}{8}$

6. $\frac{3k}{5} + 44 = \frac{12k}{25} + 8$

_____ _____ _____

7. $10a - 37 = 6a + 51$

8. $5w + 9.9 = 4.8 + 8w$

9. $15 - x = 2(x + 3)$

_____ _____ _____

10. $15y + 14 = 2(5y + 6)$

11. $14 - \frac{w}{8} = \frac{3w}{4} - 21$

12. $\frac{1}{2}(6x - 4) = 4x - 9$

_____ _____ _____

13. $4(3d - 2) = 8d - 5$

14. $\frac{y}{3} + 11 = \frac{y}{2} - 3$

15. $\frac{2x - 9}{3} = 8 - 3x$

_____ _____ _____

16. Forty-eight decreased by a number is the same as the difference of four times the number and seven. Find the number.

_____

17. The square and the equilateral triangle at the right have the same perimeter. Find the length of the sides of the triangle.

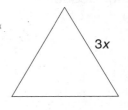

_____

Practice C
Lesson 11-3: Solving Equations with Variables on Both Sides

Solve.

1. $9x - 22 = 23 + 4x$

2. $\frac{2y}{3} - 9 = 2y + 3$

3. $5x - 36 = 24 + 2x$

4. $\frac{7n}{9} - 62 = \frac{5n}{9} - 48$

5. $\frac{x+4}{5} = \frac{x-6}{7}$

6. $8(k + 6) = 3(k + 33)$

7. $\frac{2x-1}{3} = 4x + 3$

8. $7d + 13 = 35 - 4d$

9. $\frac{r}{5} - 26 = \frac{r}{2} - 29$

10. $\frac{w}{4} + 5 = \frac{w}{3} + 10$

11. $\frac{2m}{5} + 17 = \frac{m}{2} - 7$

12. $5(5a + 3) = 14a - 29$

13. $\frac{2x-9}{3} = 10 - \frac{1}{5}x$

14. $5m - 6 = 2.5m - 42$

15. $\frac{3n-7}{-2} = 5 - 0.25n$

16. The sum of five and eight times a number is the same as fifty plus one-half the number. Find the number.

17. Jack and Jessica earned the same amount last week. They both work for the same hourly rate. Jack worked eighteen hours and had $42 deducted from his pay. Jessica worked fifteen hours and had $18 deducted from her pay. What was each person's salary last week after deductions?

Name _____ Date _____ Class _____

LESSON 11-3 Reteach
Solving Equations with Variables on Both Sides

If there are variable terms on both sides of an equation, first collect them on one side. Do this by adding or subtracting.

If possible, collect the variable terms on the side where the on coefficient will be positive.

$\begin{aligned}5x &= 2x + 12\\ -2x & -2x\end{aligned}$ To collect on left side, subtract $2x$.

Check: Substitute into the original equation.

$3x = 12$

$\dfrac{3x}{3} = \dfrac{12}{3}$ Divide by 3.

$x = 4$

$5x = 2x + 12$

$5(4) \stackrel{?}{=} 2(4) + 12$

$20 \stackrel{?}{=} 8 + 12$

$20 = 20$ ✔

$\begin{aligned}-6z + 28 &= 9z - 2\\ +6z & +6z\end{aligned}$ To collect on right side, add $6z$.

$28 = 15z - 2$

$+2 +2$ Add 2.

$30 = 15z$

$\dfrac{30}{15} = \dfrac{15z}{15}$ Divide by 15.

$2 = z$

Check: $-6z + 28 = 9z - 2$

$-6(2) + 28 \stackrel{?}{=} 9(2) - 2$

$-12 + 28 \stackrel{?}{=} 18 - 2$

$16 = 16$ ✔

Complete to solve and check each equation.

1. $9m = 4m - 25$ To collect on left, subtract.
   $- -$
   $5m = -25$
   $\dfrac{m}{\underline{}} = \dfrac{-25}{\underline{}}$ Divide.
   $m = \underline{}$

   Check: $9m = 4m - 25$
   $9(\underline{}) \stackrel{?}{=} 4(\underline{}) - 25$
   $\underline{} \stackrel{?}{=} \underline{} - 25$
   $\underline{}$

2. $3h - 7 = 5h + 1$ To collect on right, subtract.
   $-\underline{} -\underline{}$
   $-7 = \underline{}h + 1$
   $\underline{} \underline{}$ Subtract.
   $\underline{} = \underline{}h$
   $\dfrac{\underline{}}{\underline{}} = \dfrac{h}{}$ Divide.
   $\underline{} = h$

   Check: $3h - 7 = 5h + 1$
   $3(\underline{}) - 7 \stackrel{?}{=} 5(\underline{}) + 1$
   $\underline{} - 7 \stackrel{?}{=} \underline{} + 1$
   $\underline{}$

Reteach
LESSON 11-3 Equations with Variables on Both Sides (continued)

To solve multi-step equations with variables on both sides: 1) *clear* fractions, 2) *combine* like terms, 3) *collect* variable terms on one side, and 4) *isolate* the variable by using properties of equality.

$\frac{t}{3} - \frac{5t}{6} + \frac{1}{2} = t - 1$	To clear fractions, determine LCD = 6.
$6 \cdot \frac{t}{3} - 6 \cdot \frac{5t}{6} + 6 \cdot \frac{1}{2} = 6 \cdot t - 6 \cdot 1$	Multiply *every* term by the LCD.
$2t - 5t + 3 = 6t - 6$	The fractions are cleared.
$-3t + 3 = 6t - 6$	Combine like terms.
$\underline{+3t \qquad +3t}$	To collect variable terms, add $3t$ to both sides.
$3 = 9t - 6$	
$\underline{+6 \qquad +6}$	Add 6 to both sides.
$9 = 9t$	
$\frac{9}{9} = \frac{9t}{9}$	Divide both sides by 9.
$1 = t$	

Complete to solve.

3. $\frac{w}{4} + \frac{w}{2} + \frac{1}{4} = w$ Find the LCD of 2 and 4.

  ___ $\cdot \frac{w}{4}$ + ___ $\cdot \frac{w}{2}$ + ___ $\cdot \frac{1}{4}$ = ___ $\cdot w$ Multiply *every* term by the LCD.

  ___ w + ___ w + ___ = ___ w

  ___ w + ___ = ___ w Combine like terms.

  − ___   − ___   Subtract.

  ___ = w

4. $3m + 17 - m = 10 - m - 2$ **Check:** $3m + 17 - m = 10 - m - 2$

  ___ m + ___ = ___ − m $3(\underline{\quad}) + 17 - (\underline{\quad}) \stackrel{?}{=}$

  ___ +   ___ + $10 - (\underline{\quad}) - 2$

  ___ m + 17 = ___ $-9 + 17 \underline{\quad} \stackrel{?}{=} 10 \underline{\quad} - 2$

  − ___   − ___   ___ + ___ $\stackrel{?}{=}$ ___ − 2

  $m =$ ___

  $m =$ ___

Challenge
LESSON 11-3 A Handy Tool!

A **lever** is a bar that can turn about a fixed point called the **fulcrum**.

The ancient Greek mathematician Archimedes knew the power of the *lever principle*. He has been quoted as saying "Give me a place to stand and I will move the Earth."

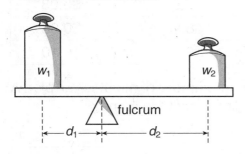

The Lever Principle

A weight w_1 is placed on one arm of a lever at a distance d_1 from the fulcrum.
A second weight w_2 is placed on the other arm at a distance d_2 from the fulcrum.

$$w_1 \cdot d_1 = w_2 \cdot d_2$$

This equation may be used to solve a problem involving the lever principle.

A 14-foot plank is used as a lever with a 120-lb box on one end and a 90-lb box on the other end. If the boxes balance one another, how far from the fulcrum is each box?

Let x = 120-lb box's distance from fulcrum.
Then $14 - x$ = 90-lb box's distance from fulcrum.

$$w_1 \cdot d_1 = w_2 \cdot d_2$$
$$120 \cdot x = 90 \cdot (14 - x)$$
$$120x = 1260 - 90x$$
$$\underline{+90x \qquad\qquad +90x}$$
$$210x = 1260$$
$$\frac{210x}{210} = \frac{1260}{210}$$

$x = 6$ ft ⟵ 120-lb box's distance from the fulcrum.
$14 - x = 8$ ft ⟵ 90-lb box's distance from the fulcrum.

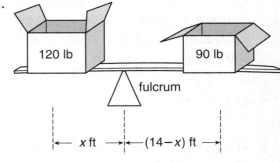

Write and solve an equation.

A 21-ft plank is used as a lever with a 108-lb barrel on one end and a 81-lb barrel on the other end. If the barrels balance one another, how far from the fulcrum is the 108-lb barrel?

The 108-lb barrel is _____ from the fulcrum.

Problem Solving
11-3 Solving Equations with Variables on Both Sides

The chart below describes three long-distance calling plans. Round to the nearest minute. Write the correct answer.

1. For what number of minutes will plan A and plan B cost the same?

   _____

Long-Distance Plans

Plan	Monthly Access Fee	Charge per minute
A	$3.95	$0.08
B	$8.95	$0.06
C	$0	$0.10

2. For what number of minutes per month will plan B and plan C cost the same?

   _____

3. For what number of minutes will plan A and plan C cost the same?

   _____

Choose the letter for the best answer.

4. Carpet Plus installs carpet for $100 plus $8 per square yard of carpet. Carpet World charges $75 for installation and $10 per square yard of carpet. Find the number of square yards of carpet for which the cost including carpet and installation is the same.

   A 1.4 yd^2 C 12.5 yd^2
   B 9.7 yd^2 D 87.5 yd^2

5. One shuttle service charges $10 for pickup and $0.10 per mile. The other shuttle service has no pickup fee but charges $0.35 per mile. Find the number of miles for which the cost of the shuttle services is the same.

   F 2.5 miles
   G 22 miles
   H 40 miles
   J 48 miles

6. Joshua can purchase tile at one store for $0.99 per tile, but he will have to rent a tile saw for $25. At another store he can buy tile for $1.50 per tile and borrow a tile saw for free. Find the number of tiles for which the cost is the same. Round to the nearest tile.

   A 10 tiles   C 25 tiles
   B 13 tiles   D 49 tiles

7. One plumber charges a fee of $75 per service call plus $15 per hour. Another plumber has no flat fee, but charges $25 per hour. Find the number of hours for which the cost of the two plumbers is the same.

   F 2.1 hours   H 7.5 hours
   G 7 hours   J 7.8 hours

Reading Strategies
11-3 Follow a Procedure

Equations may have variables on both sides. Follow these steps to get the variables on one side of the equation.

Solve $6x - 7 = 2x + 5$.

Step 1: Get all variables on one side of the equation.
$6x - 2x - 7 = 2x - 2x + 5$ Subtract **2x** from both sides.
$4x - 7 = 5$

Step 2: Get all constants on the other side of the equation.
$4x - 7 + 7 = 5 + 7$ Add **7** to both sides.
$4x = 12$

Step 3: Solve.
$\frac{4x}{4} = \frac{12}{4}$ Divide both sides by 4.
$x = 3$

Use the above procedure to answer each question.

1. What is the first step to solve equations with variables on both sides?

2. What was done to get the variables on one side?

3. Write the equation with the variables on one side only.

4. What is the second step in solving the equation?

5. What was done to get the constants on one side?

6. What was the last step to solve the equation?

Puzzles, Twisters & Teasers

Lesson 11-3: Getting a New CD!

You've just earned some extra money and you want to buy a new CD, so you need to get to the music store. Start by solving the equations below. Once you have the solutions follow the directions to work your way through the maze.

1. $5x - 2 = x + 6$ $x = $ ____ Start at the S and go right x spaces.
2. $6k - 6 = 6 + 4k$ $k = $ ____ Go down k spaces.
3. $2a + 3 = 3a - 2$ $a = $ ____ Go right a spaces.
4. $4(t - 5) + 2 = t + 3$ $t = $ ____ Go right t spaces.
5. $2c + 4 - 3c = -9 + c + 5$ $c = $ ____ Go up c spaces.
6. $5n - 3 = 2n + 12$ $n = $ ____ Go right n spaces.
7. $3d + 4 = d + 18$ $d = $ ____ Go down d spaces.

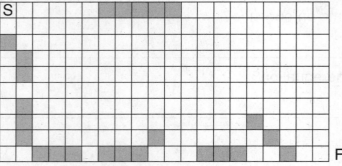

Practice A
11-4 Solving Inequalities by Multiplying or Dividing

1. $4x > -20$

   _____
   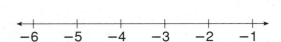

2. $3 \geq \dfrac{y}{5}$

   _____
   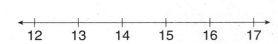

3. $-\dfrac{b}{8} \geq 3$

   _____
   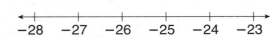

4. $-6d < 18$

   _____
   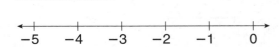

5. $63 \geq 7f$

   _____

6. $-\dfrac{g}{4} \leq 2$

   _____
   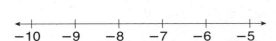

7. $13 < \dfrac{h}{3}$

   _____
   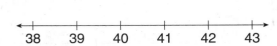

8. $-7j > -14$

   _____

9. Cheryl wants to buy a bicycle that costs $160. If she saves $12 each week, what is the fewest number of weeks she must save in order to buy the bicycle?

   _____

10. Mark worked on math homework less than $\dfrac{1}{3}$ the amount of time that his brother did. If Mark spent 25 minutes on his math homework, how much time did his brother spend on his math homework?

    _____

Name _____ Date _____ Class _____

Practice B
LESSON 11-4 Solving Inequalities by Multiplying or Dividing

Solve and graph.

1. $\dfrac{m}{-5} \le 4$

   _____
   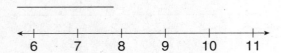

2. $-16 < -8n$

   _____
   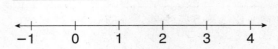

3. $7p \ge 49$

   _____
   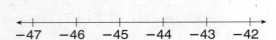

4. $10 > \dfrac{q}{2}$

   _____
   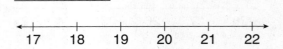

5. $-\dfrac{r}{3} \le 15$

   _____
   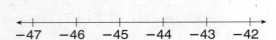

6. $22 > -2s$

   _____
   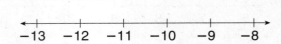

7. $-6t < -24$

   _____
   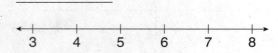

8. $\dfrac{v}{20} \ge 2$

   _____
   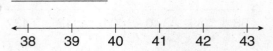

9. On a snorkeling trip, Antonia dove at least 7 times as deep as Lucy did. If Antonia dove 35 feet below the ocean's surface, what was the deepest that Lucy dove?

   _____

10. Last week, Saul ran more than one-fifth the distance that his friend Omar ran. If Saul ran 14 miles last week, how far did Omar run?

    _____

Name _____ Date _____ Class _____

LESSON 11-4 Practice C
Solving Inequalities by Multiplying or Dividing

Solve and graph.

1. $18 \geq \dfrac{b}{-3}$

   _____

   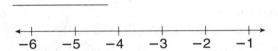

2. $6d > 42$

   _____

   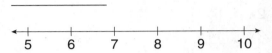

3. $5f < -15$

   _____

4. $24 \leq \dfrac{g}{2}$

   _____

   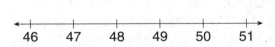

5. $-4 < \dfrac{h}{-2}$

   _____

6. $8j \geq -40$

   _____

   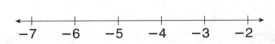

7. $\dfrac{k}{6} \geq 3$

   _____

8. $-56 < -7m$

   _____

   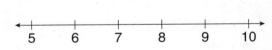

9. Keisha told her swim coach that she would swim at least half a mile. If one lap is 50 yards, what is the fewest number of laps she must swim?

   _____

10. Mr. Wallace measured the length of a board that he was going to cut into 16 equal parts. If each part had to be less than 4 inches long, how long could the board be?

    _____

Reteach
11-4 Solving Inequalities by Multiplying or Dividing

To solve an inequality, multiply and divide the same way you would solve an equation. But, if you multiply or divide by a negative number, you must reverse the inequality sign.

Divide by a Positive Number

$2x < 14$

$\dfrac{2x}{2} < \dfrac{14}{2}$

$x < 7$

To check your solution, choose two numbers from the graph and substitute them into the original equation. Choose a number that should be a solution and a number that should not be a solution.

Divide by a Negative Number

$-2x < 14$

$\dfrac{-2x}{-2} > \dfrac{14}{-2}$ Reverse the inequality sign.

$x > -7$

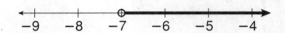

Check

According to the graph, -6 should be a solution, but -8 should not be.

$$-2x < 14 \qquad -2x < 14$$
$$\dfrac{-2 \cdot -8}{-2} \overset{?}{>} \dfrac{14}{-2} \quad \dfrac{-2 \cdot -6}{-2} \overset{?}{>} \dfrac{14}{-2}$$
$$-8 > -7 \; ✗ \qquad -6 > -7 \; ✓$$

Complete to solve. Then graph the equation and check.

1. $-3y \geq 24$

   $\dfrac{-3y}{-3}$ _____ $\dfrac{24}{-3}$

   y _____ _____

   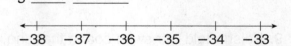

2. $\dfrac{s}{-9} < 4$

   _____ $\cdot \dfrac{s}{-9}$ _____ _____ $\cdot 4$

   s _____ _____

Challenge

11-4 Compounding the Problem

Inequalities that have more than one inequality sign are compound inequalities. You can solve them the same way you solve other inequalities. But, you must check to be sure the solution makes sense.

Example 1

$4 < 2x < 10$

$\dfrac{4}{2} < \dfrac{2x}{2} < \dfrac{10}{2}$ Divide each part by 2.

$2 < x < 5$ This means that $x > 2$ and $x < 5$. That makes sense.

Example 2

$-4 \geq \dfrac{x}{-3} \geq 2$

$-3 \cdot -4 \leq -3 \cdot \dfrac{x}{-3} \leq -3 \cdot 2$ Multiply each part by -3. Reverse the inequality symbols.

$12 \leq x \leq -6$ This means that $x \geq 12$ and $x \leq -6$. That does not make sense, so the inequality has no solution.

Solve and graph. If the inequality has no solution, write no solution.

1. $5 > \dfrac{x}{5} > 3$

2. $14 \leq 2y < 18$

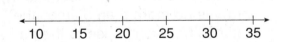

3. $20 < -5z < 35$

4. $1 \leq \dfrac{b}{-2} < -2$

Problem Solving
11-4 Solving Inequalities by Multiplying or Dividing

Write the correct answer

1. A bottle contains at least 4 times as much juice as a glass contains. The bottle contains 32 fluid ounces. Write an inequality that shows this relationship.

2. Solve the inequality in Exercise 1. What is the greatest amount the glass could contain?

3. In the triple jump, Katrina jumped less than one-third the distance that Paula jumped. Katrina jumped 5 ft 6 in. Write an inequality that shows this relationship.

4. Solve the inequality in Exercise 3. How far could Paula could have jumped?

Choose the letter for the best answer.

5. Melinda earned at least 3 times as much money this month as last month. She earned $567 this month. Which inequality shows this relationship?
   A $567 < x$
   B $567 < 3x$
   C $567 > 3x$
   D $567 \geq 3x$

6. The shallow end of a pool is less than one-quarter as deep as the deep end. The shallow end is 3 feet deep. Which inequality shows this relationship?
   F $4 > 3x$
   G $4x < 3$
   H $\frac{x}{4} > 3$
   J $\frac{x}{4} < 3$

7. Arthur worked in the garden more than half as long as his brother. Arthur worked 6 hours in the garden. How long did his brother work in the garden?
   A less than 3 hours
   B 3 hours
   C less than 12 hours
   D more than 12 hours

8. The distance from Bill's house to the library is no more than 5 times the distance from his house to the park. If Bill's house is 10 miles from the library, what is the greatest distance his house could be from the park?
   F 2 miles
   G more than 2 miles
   H 20 miles
   J less than 20 miles

Name _____ Date _____ Class _____

Reading Strategies
LESSON 11-4 Understand Symbols

If you know the meanings of the inequality symbols, you can read and write inequalities as word sentences, and you can write word sentences as inequalities.

   $<$ less than
   $>$ greater than, or more than
   $\leq$ less than or equal to, or no more than
   $\geq$ greater than or equal to, or at least

Inequality	Word Sentence
$6 < x$	Six is less than x.
$y > 14$	y is greater than fourteen, or y is more than fourteen.
$5 \leq z$	Fifteen is less than or equal to z, or fifteen is no more than z.
$b \geq 5$	b is greater than or equal to five, or b is at least 5.

Many inequalities include multiplication or division.

Inequality	Word Sentence
$21 < 3x$	Twenty-one is less than three times x.
$\frac{y}{3} > 8$	y divided by three is greater than eight.
$4 \leq 2z$	Four is less than or equal to two times z, or four is no more than $2z$.
$\frac{b}{4} \geq 10$	b divided by four is greater than or equal to ten, or b divided by four is at least ten.

Write the inequality as a word sentence.

1. $5d > 40$ _____

2. $\frac{f}{6} \leq 3$ _____

3. $11 < \frac{g}{2}$ _____

4. $16 \geq 4h$ _____

Write an inequality that you could use to solve the problem.

5. A tree is more than five times as tall as a math student. The tree is 28 feet tall.

_____

Name _____ Date _____ Class _____

Puzzles, Twisters, and Teasers
LESSON 11-4 *Other Things Being Equal*

Complete to solve each inequality. Use the key to match each answer with its letter. Unscramble the letters to answer the riddle.

Key

A	B	C	D	E	F	G	H	I	J	K	L	M
1	2	3	4	>	6	<	8	9	10	11	12	13

N	O	P	Q	R	S	T	U	V	W	X	Y	Z
14	15	16	≤	18	19	20	21	22	23	24	25	≥

1. $\dfrac{x}{-6} > -2$ → $x <$ _____

2. $\dfrac{j}{-3} < -7$ → $j >$ _____

3. $400 \leq 400y$ → $y \geq$ _____

4. $65 > -13d$ → d _____ -5

5. $\dfrac{f}{7} < 2$ → $f <$ _____

6. $17g \leq -68$ → g _____ -4

7. $3 \geq \dfrac{h}{7}$ → $h \leq$ _____

How are all numbers created?

___ ___ ___ ___ ___ ___ ___

Practice A
11-5 Solving Two-Step Inequalities

Write *yes* or *no* to tell whether the inequality symbol would be reversed in the solution. Do not solve.

1. $2x - 4 < 20$
2. $4 - 3y \leq 21$
3. $6x + 17 > 3$
4. $-\frac{a}{5} - 4 \geq -2$

Solve.

5. $2x - 17 \geq 29$
6. $8 - \frac{k}{2} < -12$
7. $23 - 3w < -34$
8. $24 - 0.6x < 60$

9. $10 \leq 5 - 2d$
10. $\frac{x}{3} + 5 \leq 14$
11. $\frac{2}{3} \geq \frac{y}{6} - \frac{1}{2}$
12. $\frac{-a}{7} + \frac{1}{7} > \frac{1}{14}$

Solve and graph.

13. $2x - 1 < 3$

14. $16 \geq 1 - 3a$

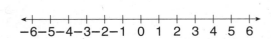
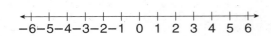

15. $\frac{y}{2} - \frac{3}{4} \leq \frac{1}{2}$

16. $\frac{d}{3} + \frac{5}{12} > \frac{1}{4}$

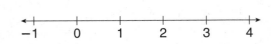
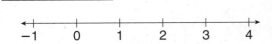

17. Mrs. Ocosta is paid a 5% commission on her sales each week. In addition, she receives a base salary of $375. What should the amount of her sales be for the week if she hopes to make at least $600 this week?

Name _____ Date _____ Class _____

Practice B
LESSON 11-5 Solving Two-Step Inequalities

Solve and graph.

1. $4x - 2 < 26$

   _____

   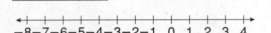

2. $6 - \frac{1}{5}y \leq 7$

   _____

   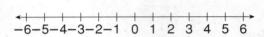

3. $2x + 27 \geq 15$

   _____

   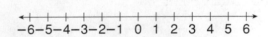

4. $10x > 14x + 8$

   _____

   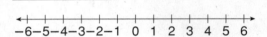

5. $7 - 4w \leq 19$

   _____

   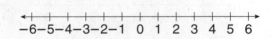

6. $\frac{k}{5} + \frac{3}{20} < \frac{3}{10}$

   _____

7. $4.8 - 9.6x \leq 14.4$

   _____

   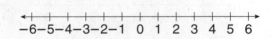

8. $\frac{2}{9} + \frac{y}{3} > \frac{1}{3}$

   _____

9. One-third of a number, decreased by thirty-six, is at most twenty-two. Find the number. _____

10. Jack wants to run at least 275 miles before the baseball season begins. He has already run 25 miles. He plans to run 2.5 miles each day. At this rate, what is the fewest number of days he will need to reach his goal? _____

Practice C
11-5 Solving Two-Step Inequalities

Solve and graph.

1. $18 - 8x > 2$

   _____
   $\leftarrow|+|+|+|+|+|+|+|+|+|+|+|+|\rightarrow$
   $-6-5-4-3-2-1\ 0\ 1\ 2\ 3\ 4\ 5\ 6$

2. $-\dfrac{5}{6} \leq -\dfrac{m}{12} - \dfrac{3}{4}$

   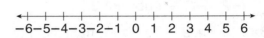

3. $-2(w - 6) > 2$

   _____
   $\leftarrow|+|+|+|+|+|+|+|+|+|+|+|+|\rightarrow$
   $-6-5-4-3-2-1\ 0\ 1\ 2\ 3\ 4\ 5\ 6$

4. $\dfrac{2(a-6)}{-9} \geq 0$

5. $9(3x - 7) < 18$

6. $\dfrac{d}{4} - 2 \geq -1$

   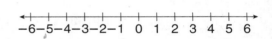

7. $\dfrac{3(x+2)}{-4} \leq 5 - 4x$

   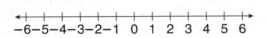

8. $\dfrac{4n - 3}{5} < \dfrac{3(n-1)}{4}$

   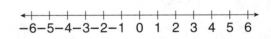

9. Deidre goes out for lunch and realizes that she only has $10 with her. If she wants to leave a 20% tip and knows she will have to pay 5% tax on her lunch, what is the most expensive lunch she can order?

   _____

10. Mr. and Mrs. Schaefer are selling their house. Their real estate agent's fee is 7% of the selling price. The Schaefers want to make at least $175,000 from the sale. To the nearest dollar, what must be the selling price of the house?

    _____

Reteach

11-5 Solving Two-Step Inequalities

To solve an inequality, undo operations the same way you would with an equation. But, when multiplying or dividing by a negative number, reverse the inequality symbol.

$3x + 2 > 11$	To undo addition,	$-3x + 2 > 11$	To undo addition,
$\underline{-2\ -2}$	subtract 2.	$\underline{-2\ -2}$	subtract 2.
$3x\ \ \ > 9$	To undo multiplication,	$-3x\ \ \ > 9$	To undo multiplication,
$\dfrac{3x}{3} > \dfrac{9}{3}$	divide by 3.	$\dfrac{-3x}{-3} < \dfrac{9}{-3}$	divide by −3 and
$x\ \ \ > 3$		$x\ \ \ < -3$	change > to <.

The solution set contains all real numbers greater than 3.

The solution set contains all real numbers less than −3.

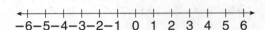

Complete to solve and graph.

1. $2t + 1 \leq 9$ To undo addition,
   _____ __   subtract.
   $2t\ \ \leq$ ___   To undo multiplication,
   $\dfrac{2t}{}\ \ \underline{}$ divide.
   $t\ \ \underline{}$

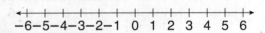

2. $-2t + 1 \leq 9$ To undo addition,
   _____ __   subtract.
   $-2t\ \ \leq 8$ To undo multiplication,
   $\dfrac{-2t}{}\ \ \underline{}$ divide by −2 and
   $t\ \ \underline{}$ change ≤ to ≥.

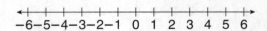

3. $-3z - 2 > 1$
   _____   _____
   $-3z >$ _____
   $\dfrac{-3z}{}\ \ \underline{}$
   $z\ \ \underline{}$

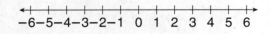

4. $3z - 2 > 1$
   _____   _____
   $3z >$ _____
   $\dfrac{3z}{}\ \ \underline{}$
   $z\ \ \underline{}$

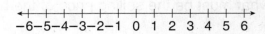

Name _____ Date _____ Class _____

Reteach
LESSON 11-5 Solving Two-Step Inequalities (continued)

To solve multistep inequalities, you may need to clear fractions.
Multiply both sides by the LCD.

$\frac{v}{4} + \frac{1}{4} > -\frac{1}{2}$ The LCD is 4.

$4 \cdot \frac{v}{4} + 4 \cdot \frac{1}{4} > 4 \cdot -\frac{1}{2}$ Multiply by the LCD.

$v + 1 > -2$

$\underline{-1} \quad \underline{-1}$ Subtract from both sides.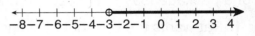

$v > -3$

Complete to solve and graph.

5. $-\frac{b}{4} - \frac{7}{12} \leq \frac{2}{3}$ Find the LCD.

_____ $\cdot -\frac{b}{4} -$ _____ $\cdot \frac{7}{12} \leq$ _____ $\cdot \frac{2}{3}$ Multiply by the LCD.

_____ $b -$ _____ $\leq$ _____

_____ _____ _____   Add.

_____ $b \leq$ _____

_____ b ___   Divide and change symbol.

b _____   Check direction.

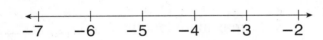

6. $\frac{2}{7} + \frac{y}{14} \geq -\frac{1}{2}$

___ $\cdot \frac{2}{7} +$ ___ $\cdot \frac{y}{14} \geq$ ___ $\cdot -\frac{1}{2}$

_____ $+$ _____ $\geq$ _____

_____   _____

_____ $\geq$ _____

7. $-\frac{1}{3} > \frac{x}{9} + \frac{2}{3}$

___ $\cdot -\frac{1}{3} >$ ___ $\cdot \frac{x}{9} +$ ___ $\cdot \frac{2}{3}$

_____ $> x +$ _____

_____   _____

_____ $> x$

x ___ _____

Name _____ Date _____ Class _____

LESSON 11-5 Challenge
Updated Pony Express

Pat wants to send some copies of her newly published book to friends.

According to the U.S. Postal Service:

Rates are based on the weight of the piece and the zone (distance from origin to destination ZIP code).

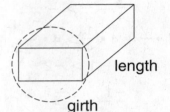

The combined length and girth (perimeter of an end) of a package may not exceed 108 inches.

1. Pat wants the box that contains books to be 6 inches high, and twice as long as it is wide.

   Let x represent the width of a box that Pat might use.

   Write and solve an inequality to find all possible widths for a box that will satisfy the postal requirements and Pat's conditions.

   possible width: _____

2. Pat's husband, Mike, suggests that the box be 8 inches high and that the length be 3 times the width.

   Let z represent the length of a box that Mike suggests.

   Write and solve an inequality to find, to the nearest inch, the maximum length for a box that will satisfy.

   maximum length: _____

3. On May 1, 2002, Pat shipped a box containing a book to a friend who lives in Zone 4. Pat paid $2.08 to ship this package.

   According to the table below, write an inequality to show the weight of this package. _____

Bound Printed Matter Rates

Weight Not Over (pounds)	Local, Zones 1&2	Zone 3	Zone 4	Zone 5	Zone 6	Zone 7	Zone 8
1.0	$1.80	$1.83	$1.87	$1.93	$1.99	$2.06	$2.21
1.5	1.80	1.83	1.87	1.93	1.99	2.06	2.21
2.0	1.84	1.88	1.94	2.02	2.10	2.19	2.38
2.5	1.90	1.95	2.00	2.11	2.21	2.33	2.57
3.0	1.94	2.00	2.08	2.20	2.32	2.46	2.75
3.5	1.99	2.06	2.15	2.29	2.43	2.60	2.93
4.0							

Problem Solving
11-5 Solving Two-Step Inequalities

A school club is selling printed T-shirts to raise $650 for a trip. The table shows the profit they will make on each shirt after they pay the cost of production.

1. Suppose the club already has $150, at least how many 50/50 shirts must they sell to make enough money for the trip?

Shirt	Profit
50/50	$5.50
100% cotton	$7.82

2. Suppose the club already has $100, but it plans to spend $50 on advertising. At least how many 100% cotton shirts must they sell to make enough money for the trip?

3. Suppose the club sold thirty 50/50 shirts on the first day of sales. At least how many more 50/50 shirts must they sell to make enough money for the trip?

For Exercises 4–5, use this equation to estimate typing speed, $S = \frac{w}{5} - 2e$, where S is the accurate typing speed, w is the number of words typed in 5 minutes, and e is the number of errors. Choose the letter for the best answer.

4. One of the qualifications for a job is a typing speed of at least 65 words per minute. If Jordan knows that she will be able to type 350 words in five minutes, what is the maximum number of errors she can make?

   A 0 C 3
   B 2 D 4

5. Tanner usually makes 3 errors every 5 minutes when he is typing. If his goal is an accurate typing speed of at least 55 words per minute, how many words does he have to be able to type in 5 minutes?

   F 61 words   H 305 words
   G 300 words J 325 words

6. A taxi charges $2.05 per ride and $0.20 for each mile, which can be written as $F = \$2.05 + \$0.20m$. How many miles can you travel in the cab and have the fare be less than $10?

   A 15 C 39
   B 25 D 43

7. Celia's long distance company charges $5.95 per month plus $0.06 per minute. If Celia has budgeted $30 for long distance, what is the maximum number of minutes she can call long distance per month?

   F 375 minutes   H 405 minutes
   G 400 minutes   J 420 minutes

Reading Strategies
LESSON 11-5 Follow a Procedure

You can use these steps to help you solve a two-step inequality.

Solve $-8 < 4x + 4$.

Step 1: Get the variable by itself on one side of the inequality.

$-8 - 4 < 4x + 4 - 4$
$-12 < 4x$

Subtract 4 from both sides.

Step 2: Solve.

$-\frac{12}{4} < \frac{4x}{4}$
$-3 < x$

Divide both sides by 4.

Step 3: Rewrite the solution so the variable comes first.

$x > -3$

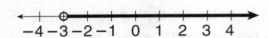

Use the procedure to answer each question.

1. What did the procedure tell you to do first?

   _____

2. How did you get the variable by itself in this problem?

   _____

3. What is the second step given?

   _____

4. How did you solve this inequality?

   _____

5. How would the graph for $x \geq -3$ be different than the above graph?

   _____

Puzzles, Twisters & Teasers

11-5 Get to the Beach!

Decide whether or not the given solution to each inequality is correct. Circle the letter above your answer. Then unscramble the letters to solve the riddle.

1. $2x - 3 > 5$ $x > 4$
   - **S** correct
   - **A** incorrect

2. $-10 < 3x + 2$ $-4 < x$
   - **A** correct
   - **U** incorrect

3. $-2x + 4 \leq 3$ $x \geq \frac{1}{2}$
   - **N** correct
   - **W** incorrect

4. $-2 - x > 5$ $x < -7$
   - **D** correct
   - **T** incorrect

5. $\frac{2x}{3} + \frac{1}{2} \leq \frac{5}{6}$ $x \leq \frac{1}{2}$
   - **W** correct
   - **A** incorrect

6. $3 > 3x - 6$ $3 > x$
   - **I** correct
   - **R** incorrect

7. $10x > 200 + 2x$ $x > 25$
   - **C** correct
   - **V** incorrect

8. $3k - 2 > 13$ $k > 3$
   - **T** correct
   - **H** incorrect

9. $10x + 2 > 42$ $x < 4$
   - **I** correct
   - **E** incorrect

10. $5 < -p - 12$ $p \geq 10$
    - **C** correct
    - **S** incorrect

What do witches like to eat at the beach?

S A N D W I C H E S

Practice A
11-6 Systems of Equations

Substitute to determine if the ordered pair (3, 2) is a solution of the following systems of equations. Write *yes* or *no*.

1. $x + y = 5$
   $2x - y = 4$
   _____

2. $y = x - 1$
   $y = 3x - 7$
   _____

3. $y = -x + 5$
   $x - 2y = -4$
   _____

4. $4x - y = 10$
   $3y = x + 1$
   _____

Solve each system of equations.

5. $y = x$
   $y = -x$

6. $y = 3x + 1$
   $y = 2x - 1$

7. $y = 2x - 3$
   $y = x - 2$

8. $y = -2x + 3$
   $y = 3x + 3$

9. $x + y = 2$
   $2x + y = 1$

10. $2x = y$
    $4x - y = -2$

11. $2x + 4y = 24$
    $x - 2y = -8$

12. $3x + y = 4$
    $6x + 2y = 8$

13. The sum of two numbers is 8. The difference of the two numbers is 2. Write a system of equations to find these two numbers and solve.

Name _____ Date _____ Class _____

Practice B
LESSON 11-6 Systems of Equations

Solve each system of equations.

1. $y = 2x - 4$
   $y = x - 1$

   _____

2. $y = -x + 10$
   $y = x + 2$

   _____

3. $y = 2x - 1$
   $y = -3x - 6$

   _____

4. $y = 2x$
   $y = 12 - x$

   _____

5. $y = 2x - 3$
   $y = 2x + 1$

   _____

6. $y = 3x - 1$
   $y = x + 1$

   _____

7. $x + y = 0$
   $5x + 2y = -3$

   _____

8. $2x - 3y = 0$
   $2x + y = 8$

   _____

9. $2x + 3y = 6$
   $4x + 6y = 12$

   _____

10. $6x - y = -14$
    $2x - 3y = 6$

    _____

11. The sum of two numbers is 24. The second number is 6 less than the first. Write a system of equations and solve it find the number.

    _____

15. Kerry and Luke biked a total of 18 miles in one weekend. Kerry biked 4 miles more than Luke. Write a system of equations and solve it to find how far each boy biked.

    _____

    _____

Holt Mathematics

Name _____ Date _____ Class _____

LESSON 11-6
Practice C
Systems of Equations

Solve each system of equations.

1. $y = 3x - 1$
   $y = -2x$

2. $2x + 3y = 12$
   $x = 4y - 5$

3. $-x + 2y = 8$
   $4x + y = -5$

4. $2x - y = 7$
   $3x - 4y = 8$

5. $y = -0.4x$
   $y = -0.8x + 2$

6. $2x - y = 4$
   $\frac{1}{2}y = x - 2$

7. $2y = 5x + 5$
   $7y - 3x = 32$

8. $y = \frac{1}{4}x - 1$
   $x + 4y = -4$

9. $2x + 3y = 3$
   $3y = 5 - 2x$

10. $4x - y = 3$
    $4x + y = 1$

11. The sum of two numbers is 381. Their difference is 155. Find the two numbers.

12. The theater group sold adult and student tickets for the play. They sold a total of 560 tickets. Each adult ticket was $8 and each student ticket was $3.50. The group took in a total of $3166. How many of each type of ticket were sold for the play?

13. The cost of 20 blank recordable CDs and 6 music CDs is $80.70. The cost of 30 blank recordable CDs and 4 music CDs is $66.30. Find the cost of each blank recordable CD and each music CD.

LESSON 11-6 Reteach
Systems of Equations

Two or more equations considered together form a **system of equations**. To solve a system of equations, you can use a method of substitution.

Solve the system: $y = 3x$
$y - 5x = 20$

Use the first equation to substitute for y in the second equation.	$y - 5x = 20$ $3x - 5x = 20$	second equation Replace y with $3x$.
Solve the resulting equation for x.	$-2x = 20$ $\frac{-2x}{-2} = \frac{20}{-2}$ $x = -10$	Combine like terms. Divide by -2.
Substitute the x-value into the first equation to get the corresponding y-value.	$y = 3x$ $y = 3(-10) = -30$	

Check: Substitute both values in each of the original equations.

$y = 3x$ $y - 5x = 20$
$-30 \stackrel{?}{=} 3(-10)$ $-30 - 5(-10) \stackrel{?}{=} 20$ $x = -10$ and $y = -30$
$-30 = -30$ ✓ $-30 + 50 \stackrel{?}{=} 20$ So, the solution of the
              $20 = 20$ ✓   system is $(-10, -30)$.

Solve and check this system.

1. $y = 2x$
   $6x + y = 16$

   Use the first equation to substitute for y in the second equation.

   $6x + \underline{\quad} = 16$

   Solve the resulting equation for x.

   _____

   _____

   Substitute the x-value to get the corresponding y-value.

   $y = 2x$

   $y = 2(\underline{\quad}) = \underline{\quad}$

   Check both values in each of the original equations.

   $y = 2x$ $6x + y = 16$

   $\underline{\quad} \stackrel{?}{=} 2(\underline{\quad})$ $6(\underline{\quad}) + \underline{\quad} \stackrel{?}{=} 16$

   _____   _____ $\stackrel{?}{=} 16$

                      _____

   So, the ordered pair _____ is the solution of the system.

Name _____ Date _____ Class _____

Reteach
LESSON 11-6 Systems of Equations (continued)

Sometimes, you first have to solve one equation for a variable.
Solve the system: $y + 3x = 7$
$x + 2y = 4$

Solve the first equation for y.	$y + 3x = 7$	
	$\underline{-3x -3x}$	Subtract $3x$.
	$y = 7 - 3x$	

Substitute for y in the second equation.	$x + 2y = 4$	second equation
	$x + 2(7 - 3x) = 4$	Replace y with $7 - 3x$.
Solve for x.	$x + 14 - 6x = 4$	Distributive property
	$-5x + 14 = 4$	Combine like terms.
	$\underline{-14 -14}$	Subtract 14.
	$\dfrac{-5x}{-5} = \dfrac{-10}{-5}$	Divide by -5.
	$x = 2$	

Substitute the x-value into the first equation to get the corresponding y-value.	$y + 3x = 7$
	$y + 3(2) = 7$
	$y + 6 = 7$
	$y + 6 - 6 = 7 - 6$
	$y = 1$

Check: Substitute both values in each of the original equations.

$y + 3x = 7$	$x + 2y = 4$	$x = 2$ and $y = 1$
$1 + 3(2) \stackrel{?}{=} 7$	$2 + 2(1) \stackrel{?}{=} 4$	The solution of the system is $(2, 1)$.
$7 = 7$ ✓	$4 = 4$ ✓	

Solve and check this system.

2. $y - 2x = 0$
$x - 2y = 6$

Solve the first equation for y.
$y - 2x = 0$

$y =$ _____

Use the result to substitute for y in the second equation.
$x - 2y = 6$
$x - 2(\underline{}) = 6$

Solve the resulting equation for x.

$x =$ _____

Substitute the x-value to get the corresponding y-value.

_____

_____

$y =$ _____

Check:
$y - 2x = 0 x - 2y = 6$

So, the ordered pair _____ is the solution of the system.

LESSON 11-6 Challenge
Different Strokes!

The first objective in solving a system of two equations in two variables is to get to one equation in one variable. You have seen how to do this by using substitution.

Now you will see how to solve the system at the right by using addition.

$x + y = 7$
$2x - y = 2$

$$\begin{array}{r} x + y = 7 \\ 2x - y = 2 \\ \hline 3x = 9 \end{array}$$

Add the equations to eliminate y.
Solve for x.

$\dfrac{3x}{3} = \dfrac{9}{3}$
$x = 3$

Return to any equation with x and y.
Substitute the x-value.
Solve for y.

$x + y = 7$
$3 + y = 7$
$\underline{-3 -3}$
$y = 4$

You should check the values by substituting into the original equations.

The solution for this system is (3, 4).

If the coefficients of one variable in a set of equations are equal but not opposite, you can use subtraction to eliminate that variable.

Use the space provided to solve each system using addition or subtraction to eliminate a variable. Check your results.

1. $2x + 3y = 7$
   $x - 3y = 8$

2. $x + 3y = 7$
   $-x + 2y = 8$

3. $x - y = 9$
   $x + 4y = 24$

Solution: _____   Solution: _____   Solution: _____

Name _____ Date _____ Class _____

Problem Solving
LESSON 11-6 Systems of Equations

After college, Julia is offered two different jobs. The table summarizes the pay offered with each job. Write the correct answer.

1. Write an equation that shows the pay y of Job A after x years.

   _____

Job	Yearly Salary	Yearly Increase
A	$20,000	$2500
B	$25,000	$2000

2. Write an equation that shows the pay y of Job B after x years.

   _____

3. Is (8, 35,000) a solution to the system of equations in Exercises 1 and 2?

   _____

4. Solve the system of equations in Exercises 1 and 2.

5. If Julia plans to stay at this job only a few years and pay is the only consideration, which job should she choose?

A travel agency is offering two Orlando trip plans that include hotel accommodations and pairs of tickets to theme parks. Use the table below. Choose the letter for the best answer.

6. Find an equation about trip A where x represents the hotel cost per night and y represents the cost per pair of theme park tickets.

   A $5x + 2y = 415$ C $8x + 6y = 415$
   B $2x + 3y = 415$ D $3x + 2y = 415$

Trip	Number of nights	Pairs of theme park tickets	Cost
A	3	2	$415
B	5	4	$725

7. Find an equation about trip B where x represents the hotel cost per night and y represents the cost per pair of theme park tickets.

   F $5x + 4y = 725$
   G $4x + 5y = 725$
   H $8x + 6y = 725$
   J $3x + 4y = 725$

8. Solve the system of equations to find the nightly hotel cost and the cost for each pair of theme park tickets.

   A ($50, $105)
   B ($125 $20)
   C ($105, $50)
   D ($115, $35)

Name _____ Date _____ Class _____

Reading Strategies
LESSON 11-6 Focus on Vocabulary

A system of equations is a set of equations with a common solution. Think of the solar system. All of the planets have the Sun in common.

Equation A: $y = 5x - 3$ ← Both equations have the
Equation B: $y = 3x + 1$ same variables, x and y.

The **common solution** of a system of equations that contains two variables is an ordered pair (x, y) that solves both equations.

(x, y)
↓ ↓
$(2, 7)$ is a solution for $y = 5x - 3$.

Use the ordered pair (2, 7) to answer the following questions.

1. What is a system of equations?

   _____

2. What are the variables in the set of equations above?

   _____

3. Which number in the solution stands for x?

   _____

4. Which number in the solution stands for y?

   _____

5. Rewrite Equation A by substituting 2 and 7 for x and y.

   _____

6. Why is (2, 7) a solution of $y = 5x - 3$?

   _____

7. Rewrite Equation B by substituting 2 and 7 for x and y.

   _____

8. Is (2, 7) a solution for Equation B? Why or why not?

   _____

9. Is (2, 7) a solution for this system of equations? Why or why not?

   _____

Puzzles, Twisters & Teasers

LESSON 11-6 *System Solution!*

Solve the crossword puzzle.

Across

1. A system of ___ is a set of two or more equations that contain two or more variables.
3. Solutions to a system of equations can be written as ___ pairs.
5. When solving systems of equations, you should find values for all the ___.
7. To ___ is to replace a variable with a value.
8. To find one variable, substitute a ___ for the other variable.

Down

2. A ___ of a system of equations is a set of values that are solutions of all the equations.
4. An ordered ___ may or may not be a solution of a system of equations.
6. It is easiest to ___ for a variable that has a coefficient of one.
7. To solve a general ___ of equations with two variables, solve both equations for one of the variables.

LESSON 11-1 Practice A
Simplifying Algebraic Expressions

Combine like terms.

1. $6s - 4s$ $2s$
2. $3k + 3k$ $6k$
3. $9b - 5b$ $4b$
4. $10x - 3x$ $7x$
5. $9b + 12b$ $21b$
6. $8m + 3 - m$ $7m + 3$
7. $11d - 6d + 4$ $5d + 4$
8. $7r + 9r - 3$ $16r - 3$
9. $15q - 8q - 6p$ $7q - 6p$
10. $3y + x - 2y$ $x + y$
11. $9h - 3h + 6g$ $6h + 6g$
12. $7a - 4a + 2b + 3b$ $3a + 5b$

Simplify.

13. $2(x + 3) + x$ $3x + 6$
14. $5(1 + y) - 3y$ $5 + 2y$
15. $3(a + 3) + 1$ $3a + 10$

Solve.

16. $3a + a = 16$ $a = 4$
17. $9x - 3x = 30$ $x = 5$
18. $5w + 3w = 24$ $w = 3$

19. Last Saturday Nakesha rented 4 movies from the video store. This Saturday she rented 2 movies. Let x represent the cost of renting each movie. Write and simplify an expression for how much more Nakesha spent last week renting movies.

     $4x - 2x;\ 2x$

20. If it costs $4 to rent each movie, how much more money did Nakesha spend last week renting movies?

     $8

LESSON 11-1 Practice B
Simplifying Algebraic Expressions

Combine like terms.

1. $8a - 5a$ $3a$
2. $12g + 7g$ $19g$
3. $4a + 7a + 6$ $11a + 6$
4. $6x + 3y + 5x$ $11x + 3y$
5. $10k - 3k + 5h$ $7k + 5h$
6. $3p - 7q + 14p$ $17p - 7q$
7. $3k + 7k + 5k$ $15k$
8. $5c + 12d - 6$ $5c + 12d - 6$
9. $13 + 4b + 6b - 5$ $8 + 10b$
10. $4f + 6 + 7f - 2$ $11f + 4$
11. $x + y + 3x + 7y$ $4x + 8y$
12. $9n + 13 - 8n - 6$ $n + 7$

Simplify.

13. $4(x + 3) - 5$ $4x + 7$
14. $6(7 + x) + 5x$ $42 + 11x$
15. $3(5 + 3x) - 4x$ $15 + 5x$

Solve.

16. $6y + 2y = 16$ $y = 2$
17. $14b - 9b = 35$ $b = 7$
18. $3q + 9q = 48$ $q = 4$

19. Gregg has q quarters and p pennies. His brother has 4 times as many quarters and 8 times as many pennies as Gregg has. Write the sum of the number of coins they have, and then combine like terms.

     $q + p + 4q + 8p;\ 5q + 9p$

20. If Gregg has 6 quarters and 15 pennies, how many total coins do Gregg and his brother have?

     165 coins

LESSON 11-1 Practice C
Simplifying Algebraic Expressions

Combine like terms.

1. $7x + 3x + 5x$ $15x$
2. $9a + 4a - 8a$ $5a$
3. $12g + 10h - 3g + h$ $9g + 11h$
4. $5b + 7 + 7b - 6$ $12b + 1$
5. $3p + 6q - p - 3q$ $2p + 3q$
6. $11s - 6s + 9 + 3s$ $8s + 9$
7. $9 + 3f + 8f - 6 - 3f$ $3 + 8f$
8. $v + 4y - 2y + 5y$ $v + 7y$
9. $15k + 8m - 7k + 2m$ $8k + 10m$
10. $2 + 6a + 9b - 3b - a$ $2 + 5a + 6b$
11. $11j + 8m - 6n + 3$ $11j + 8m - 6n + 3$
12. $13x + 9y - 6x - 8y + 1$ $7x + y + 1$

Simplify.

13. $5(y + 6) + 4$ $5y + 34$
14. $6(3x - 2) + 4x$ $22x - 12$
15. $3(2 + x) + 6x - 4$ $2 + 9x$

Solve.

16. $3x + 8 + 5x = 48$ $x = 5$
17. $4(g + 7) = 64$ $g = 9$
18. $9a - 6 = 3a + 4 + 4a$ $a = 5$

19. Suppose the lengths of the four sides of a quadrilateral are represented by the expressions $3a$, $a + 2$, $2a - 1$, and $3a + 6$. Write the sum of the lengths (the perimeter), and then simplify.

     $3a + a + 2 + 2a - 1 + 3a + 6;\ 9a + 7$

20. Find the perimeter of the quadrilateral in Exercise 19 for $a = 4$ in.

     43 in.

LESSON 11-1 Reteach
Simplifying Algebraic Expressions

The parts of an expression separated by plus or minus signs are called **terms**.

The expression shown has four terms. You can combine two of these terms to **simplify** the expression.

$$5a + 7b - 3a + 6a^2$$
$$5a - 3a + 7b + 6a^2$$
$$2a + 7b + 6a^2$$

$5a + 7b - 3a + 6a^2$

Like terms have the same variable raised to the same power.

Equivalent expressions have the same value for all values of the variables.

Some algebraic equations can be solved by first combining like terms.
Solve $4w - w = 24$.

$4w - w = 24$ Identify like terms; w is $1w$.
$3w = 24$ Combine coefficients of like terms.
$\frac{3w}{3} = \frac{24}{3}$ Divide both sides by 3.
$w = 8$

Complete to combine like terms.

1. $9z + 4z$
   $(9 + \underline{4})z$
   $\underline{13}\,z$

2. $9r + 5q - 2r$
   $(9 - \underline{2})r + 5q$
   $\underline{7}\,r + 5q$

3. $5t + 12f - t - 3f$
   $(5 - \underline{1})t + (12 - \underline{3})f$
   $\underline{4}\,t + \underline{9}\,f$

Simplify.

4. $7m + 3n - m + 2n$ $6m + 5n$
5. $15r + 4 - 3r$ $12r + 2$
6. $6x + 3z - y$ $6x + 3z - y$

Complete to solve.

7. $5h + 2h = 21$
   $\underline{7}\,h = 21$
   $\frac{7h}{7} = \frac{21}{7}$
   $h = \underline{3}$

8. $16w - 5w = 44$
   $\underline{11}\,w = 44$
   $\frac{11w}{11} = \frac{44}{11}$
   $w = \underline{4}$

9. $48 = 13x - x$
   $48 = \underline{12}\,x$
   $\frac{48}{12} = \frac{12x}{12}$
   $\underline{4} = x$

LESSON 11-1 Challenge: Mission Operation

You can create your own operation. Take a symbol, such as ▲, and tell what that symbol means using the standard operations of $+, -, \times,$ and $\div$.

Example
If $x \blacktriangle y = x + 2xy$, find $5 \blacktriangle 3$.

$x \blacktriangle y = x + 2xy$	Use the operation as defined.
$5 \blacktriangle 3 = 5 + 2(5)(3)$	Substitute 5 for x and 3 for y.
$= 5 + 30$	Carry out the standard operations.
$= 35$	

Apply the definition of ▲. First, show how to substitute the values or expressions that replace x and y. Then, carry out the standard operations to simplify completely.

1. $7 \blacktriangle 2$
   $7 + 2(7)(2)$
   35

2. $2 \blacktriangle 7$
   $2 + 2(2)(7)$
   30

3. $(a \blacktriangle b) + (b \blacktriangle a)$
   $a + 2ab + b + 2ba$
   $a + 4ab + b$

4. $[a \blacktriangle (4b)] + [(2a) \blacktriangle b]$
   $a + 2a(4b) + 2a + 2(2a)(b)$
   $3a + 12ab$

Let $a ✸ b = 2a + \frac{b}{2}$. Apply the definition of ✸. Simplify the resulting expression.

5. $(9 ✸ 4) + (3 ✸ 8)$
   $2(9) + \frac{4}{2} + 2(3) + \frac{8}{2}$
   30

6. $(r ✸ 4t) + (4r ✸ 2t)$
   $(2r + \frac{4t}{2}) + 2(4r) + \frac{2t}{2}$
   $10r + 3t$

LESSON 11-1 Problem Solving: Simplifying Algebraic Expressions

Write the correct answer.

1. An item costs x dollars. The tax rate is 5% of the cost of the item, or $0.05x$. Write and simplify an expression to find the total cost of the item with tax.
   $x + 0.05x; 1.05x$

2. A sweater costs d dollars at regular price. The sweater is reduced by 20%, or $0.2d$. Write and simplify an expression to find the cost of the sweater before tax.
   $d - 0.2d; 0.8d$

3. Consecutive integers are integers that differ by one. You can represent consecutive integers as $x, x+1, x+2$ and so on. Write an equation and solve to find three consecutive integers whose sum is 33.
   10, 11, 12

4. Consecutive even integers can be represented by $x, x+2, x+4$ and so on. Write an equation and solve to find three consecutive even integers whose sum is 54.
   16, 18, 20

Choose the letter for the best answer.

5. In Super Bowl XXXV, the total number of points scored was 41. The winning team outscored the losing team by 27 points. What was the final score of the game?
   A 33 to 8
   B 34 to 7
   C 22 to 2
   D 18 to 6

6. A high school basketball court is 34 feet longer than it is wide. If the perimeter of the court is 268, what are the dimensions of the court?
   F 234 ft by 34 ft
   G 67 ft by 67 ft
   H 70 ft by 36 ft
   J 84 ft by 50 ft

7. Julia ordered 2 hamburgers and Steven ordered 3 hamburgers. If their total bill before tax was $7.50, how much did each hamburger cost?
   A $1.50
   B $1.25
   C $1.15
   D $1.02

8. On three tests, a student scored a total of 258 points. If the student improved his performance on each test by 5 points, what was the score on each test?
   F 81, 86, 91
   G 80, 85, 90
   H 75, 80, 85
   J 70, 75, 80

LESSON 11-1 Reading Strategies: Organization Patterns

A statement written with numbers and words, such as

   3 apples + 2 pears + 4 bananas + 2 apples + 6 bananas

can be rewritten with numbers and variables, like

   $3a + 2p + 4b + 2a + 6b$.

Like variables can then be combined.

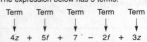

   $5a + 2p + 10b$

In an expression, **terms** are separated by + and − signs. The expression below has 5 terms:

Term Term Term Term Term
$4z$ + $5f$ + 7 − $2f$ + $3z$

Reorganize the terms so that like terms are together.
$(4z + 3z) + (5f - 2f) + 7$
Combine like terms.
$7z + 3f + 7$

Answer each question.

1. Rewrite this statement with numbers and variables:
   3 kites + 4 bats + 2 kites + 3 bats.
   $3k + 4b + 2k + 3b$

2. How many terms are in the statement above?
   4 terms

3. Reorganize terms so like terms are next to each other.
   $3k + 2k + 4b + 3b$

4. Combine like terms.
   $5k + 7b$

LESSON 11-1 Puzzles, Twisters & Teasers: Buried Treasure

Combine like terms.
Shade in the answers on the grid to reveal a hidden picture.
Hint: The picture is a drawing on a map made by a famous pirate.

1. $6x - 2x$ **$4x$**
2. $14p - 8 - 5p$ **$9p - 8$**
3. $4n + 5n + 7$ **$9n + 7$**
4. $2h + 3n + 3h + 5n + 6$ **$5h + 8n + 6$**
5. $9x + 8y + 2x + 3y - 8$ **$11x + 11y - 8$**

Simplify.
Shade the answers on the grid to continue to reveal the picture.

6. $8y - 7y$ **y**
7. $4x + 7 - 3x$ **$x + 7$**
8. $13y - 2y + 2$ **$11y + 2$**
9. $5x + 3x + 4$ **$8x + 4$**
10. $2x - 6 + 3x$ **$5x - 6$**
11. $12y - 4y$ **$8y$**
12. $5x + 2x + 3$ **$7x + 3$**
13. $4d + 2d - 3$ **$6d - 3$**

$9p - 8$	$10x$	$5z + 17$	$8f + 2g - 14$	$12y - 16$	$7g + 5h - 12$	$9n + 7$
$6y - 5t + 16$	y	$4a + 8$	$3x - 2$	$5p$	$x + 7$	$17g$
$17h - 8g + 16$	$7z + 8b - 7$	$11y + 2$	$6y + 8$	$8y$	$5a + 2x$	$7x - 9$
$5x - 3$	$19x - 4y + 8$	$18y + 3$	$6d - 3$	$14c + 18$	$7x + 6y - 8$	$4a + 8$
$10x - 2$	$15v + 6y + 3$	$8x + 4$	$9h - 7n + 5$	$5h + 8n + 6$	$12x - 4$	$34b + 5$
$3d - 6h + 14$	$11x + 11y - 8$	$3c + 4n - 2$	$6d + 4e - 12$	$9a + 6v - 3$	$5x - 6$	$12c + 13$
$4x$	$9n + 5z - 23$	$8x - 3$	$5x + 48$	$16y - 4n + 5$	$7t + 56 - 3g$	$7x + 3$

LESSON 11-2 Practice A
Solving Multi-Step Equations

Describe the operations used to solve the equation.

1. $4x + 6 - 2x = 14$
   $2x + 6 = 14$ — Combine like terms
   $2x + 6 - 6 = 14 - 6$ — Subtract 6 from both sides.
   $2x = 8$
   $\frac{2x}{2} = \frac{8}{2}$ — Divide both sides by 2.
   $x = 4$

2. $\frac{5x}{6} + \frac{2x}{3} = 3$
   $6\left(\frac{5x}{6} + \frac{2x}{3}\right) = 6(3)$ — Multiply both sides by 6.
   $5x + 4x = 18$
   $9x = 18$ — Combine like terms.
   $\frac{9x}{9} = \frac{18}{9}$ — Divide both sides by 9.
   $x = 2$

Solve.

3. $2x - 8 + 3x = 7$
   $x = 3$

4. $6y + 9 - 4y = -3$
   $y = -6$

5. $5a - 1 + a = 11$
   $a = 2$

6. $3w - 3 + 2w = -18$
   $w = -3$

7. $8 - 2d + 4d = 12$
   $d = 2$

8. $-3 = 4x + 6 - x$
   $x = -3$

9. $\frac{1}{4} = \frac{r}{4} - \frac{7}{4} + \frac{3r}{4}$
   $r = 2$

10. $\frac{2x}{3} + \frac{1}{3} - \frac{x}{3} = \frac{2}{3}$
    $x = 1$

11. $\frac{3n}{5} + \frac{9}{10} + \frac{n}{5} = -2\frac{3}{10}$
    $n = -4$

12. Together Marc and Paoli have $32. Marc has $4 less than twice the amount Paoli has. How much does each man have?
    Paoli has $12, and Marc has $20.

LESSON 11-2 Practice B
Solving Multi-Step Equations

Solve.

1. $2x + 5x + 4 = 25$
   $x = 3$

2. $9 + 3y - 2y = 14$
   $y = 5$

3. $16 = 4w + 2w - 2$
   $w = 3$

4. $26 = 3b - 2 - 7b$
   $b = -7$

5. $31 + 4t - t = 40$
   $t = 3$

6. $14 - 2x + 4x = 20$
   $x = 3$

7. $\frac{5m}{8} - \frac{6}{8} + \frac{3m}{8} = \frac{2}{8}$
   $m = 1$

8. $-4\frac{2}{3} = \frac{2n}{3} + \frac{1}{3} + \frac{n}{3}$
   $n = -5$

9. $7a + 16 - 3a = -4$
   $a = -5$

10. $\frac{x}{2} + 1 + \frac{3x}{4} = -9$
    $x = -8$

11. $7m + 3 - 4m = -9$
    $m = -4$

12. $\frac{2x}{3} + 3 - \frac{4x}{5} = \frac{1}{5}$
    $x = 7$

13. $\frac{7k}{8} - \frac{3}{4} - \frac{5k}{16} = \frac{3}{8}$
    $k = 2$

14. $6y + 9 - 4y = -3$
    $y = -6$

15. $\frac{5a}{6} - \frac{7}{12} + \frac{3a}{4} = -2\frac{1}{6}$
    $a = -1$

16. The measure of an angle is 28° greater than its complement. Find the measure of each angle.
    angle = 59°; complement = 31°

17. The measure of an angle is 21° more than twice its supplement. Find the measure of each angle.
    angle = 127°; supplement = 53°

18. The perimeter of the triangle is 126 units. Find the measure of each side.
    AC = 25 units; BC = 50 units;
    AB = 51 units

19. The base angles of an isosceles triangle are congruent. If the measure of each of the base angles is twice the measure of the third angle, find the measure of all three angles.
    36°; 72°; 72°

LESSON 11-2 Practice C
Solving Multi-Step Equations

Solve.

1. $54 + 8x + 7x = 21$
   $x = -2.2$

2. $\frac{2y}{3} + 27 + \frac{3y}{4} = -24$
   $y = -36$

3. $18x - 21 - 15x = 3$
   $x = 8$

4. $1.15a + 8 - 0.4a = -7$
   $a = -20$

5. $3 = 13w - 9w - 8$
   $w = \frac{11}{4}$ or $2\frac{3}{4}$

6. $\frac{4m}{9} + 13 - \frac{1m}{3} = 4$
   $m = -81$

7. $182d + 3.5 - 20d = 19.7$
   $d = 0.1$

8. $\frac{3k}{4} - 28 - \frac{2k}{3} = 37$
   $k = 780$

9. $\frac{3x}{4} - \frac{2x}{3} = \frac{2}{5}$
   $x = 4\frac{4}{5}$

10. Takeo and Manuel drove to a baseball game together. The tickets cost $10.50 each and parking was $6. Popcorn cost $2.25 a box and each of the boys bought a box of popcorn. Takeo bought 1 drink and Manuel bought 2 drinks. The baseball game and the snacks cost the boys a total of $39.75. How much did each of the drinks cost?
    $2.75

11. The perimeter of right triangle ABC is 132 units. Find the length of the hypotenuse.
    61 units

12. Mrs. Lincoln budgets her monthly take-home pay. The remainder she deposits in her savings account. She budgets $\frac{1}{3}$ of her take-home pay for rent, $\frac{1}{8}$ for car expenses, $\frac{1}{6}$ for food, $\frac{1}{24}$ for insurance and $\frac{1}{16}$ for entertainment. What is Mrs. Lincoln's monthly take-home pay if she deposits $546 in her savings account?
    $2016

LESSON 11-2 Reteach
Solving Multi-Step Equations

To combine like terms, add (or subtract) coefficients.
$2m + 3m = (2 + 3)m = 5m$ $x - 3x = (1 - 3)x = -2x$

To solve an equation that contains like terms, first combine the like terms.

$2m + 3m = 35 - 25$
$5m = 10$ Combine like terms.
$\frac{5m}{5} = \frac{10}{5}$ Divide by 5.
$m = 2$

Check: Substitute into the original.
$2m + 3m = 35 - 25$
$2(2) + 3(2) \stackrel{?}{=} 35 - 25$
$4 + 6 \stackrel{?}{=} 10$
$10 = 10$ ✓

$x + 6 - 3x + 5 = 13$
$-2x + 11 = 13$ Combine like terms.
$\underline{-11} \quad \underline{-11}$ Subtract 11.
$-2x = 2$
$\frac{-2x}{-2} = \frac{2}{-2}$ Divide by -2.
$x = -1$

Check: $x + 6 - 3x + 5 = 13$
$-1 + 6 - 3(-1) + 5 \stackrel{?}{=} 13$
$-1 + 6 + 3 + 5 \stackrel{?}{=} 13$
$-1 + 14 \stackrel{?}{=} 13$
$13 = 13$ ✓

Complete to solve and check each equation.

1. $4z - 7z = -20 - 1$
   $\underline{-3}z = \underline{-21}$ Combine like terms.
   $\frac{-3z}{-3} = \frac{-21}{-3}$ Divide.
   $z = \underline{7}$

   Check: $4z - 7z = -20 - 1$
   $4(\underline{7}) - 7(\underline{7}) \stackrel{?}{=} -20 - 1$
   $\underline{28} - \underline{49} \stackrel{?}{=} -21$
   $-21 = -21$ ✓

2. $t + 1 - 4t + 8 = 21$
   $\underline{-3}t + 9 = \underline{21}$ Combine like terms.
   $\underline{-9} \quad \underline{-9}$ Subtract.
   $\underline{-3}t = \underline{12}$
   $\frac{-3t}{-3} = \frac{12}{-3}$ Divide.
   $t = \underline{-4}$

   Check: $t + 1 - 4t + 8 = 21$
   $\underline{-4} + 1 - 4(\underline{-4}) + 8 \stackrel{?}{=} \underline{21}$
   $\underline{-4} + 1 + \underline{16} + 8 \stackrel{?}{=} 21$
   $\underline{-4} + \underline{25} \stackrel{?}{=} 21$
   $21 = 21$ ✓

Reteach
11-2 Solving Multi-Step Equations (continued)

To clear fractions in an equation, multiply every term by the least common denominator (LCD).

$\frac{x}{4} + 3 = \frac{1}{2}$ The LCD of 4 and 2 is 4.   Check: $\frac{1}{4}(-10) + 3 \stackrel{?}{=} \frac{1}{2}$

$4 \cdot \frac{x}{4} + 4 \cdot 3 = 4 \cdot \frac{1}{2}$ Multiply every term by LCD. $\frac{-10}{4} + \frac{12}{4} \stackrel{?}{=} \frac{1}{2}$

$x + 12 = 2$ $\frac{2}{4} \stackrel{?}{=} \frac{1}{2}$

$\underline{-12 \quad -12}$ Subtract 12. $\frac{1}{2} = \frac{1}{2}$ ✓

$x = -10$

Complete to solve and check.

3. $\frac{t}{2} + \frac{t}{6} = 2$ Determine the LCD.   Check: $\frac{1}{2}(\underline{3}) + \frac{1}{6}(\underline{3}) \stackrel{?}{=} 2$

$\underline{6} \cdot \frac{t}{2} + \underline{6} \cdot \frac{t}{6} = \underline{6} \cdot 2$ Multiply every term by LCD. $\frac{3}{2} + \frac{1}{2} \stackrel{?}{=} 2$

$\underline{3}t + t = \underline{12}$ The fractions are cleared. $\frac{4}{2} \stackrel{?}{=} 2$

$\underline{4}t = \underline{12}$ Combine like terms. $2 = 2$ ✓

$\frac{4t}{4} = \frac{12}{4}$ Divide.

$t = \underline{3}$

4. $\frac{m}{3} + \frac{m}{4} - 1 = \frac{5}{2}$ Determine the LCD.

$\underline{12} \cdot \frac{m}{3} + \underline{12} \cdot \frac{m}{4} - \underline{12} \cdot 1 = \underline{12} \cdot \frac{5}{2}$ Multiply every term by the LCD.

$\underline{4}m + \underline{3}m - \underline{12} = \underline{30}$ The fractions are cleared.

$7m - \underline{12} = \underline{30}$ Combine like terms.

$\underline{+12 \quad +12}$ Add.

$\underline{7}m = \underline{42}$

$\frac{7m}{7} = \frac{42}{7}$ Divide.

$m = \underline{6}$ Check: $\frac{6}{3} + \frac{6}{4} - 1 \stackrel{?}{=} \frac{5}{2}$

$\frac{24}{12} + \frac{18}{12} - \frac{12}{12} \stackrel{?}{=} \frac{30}{12}$

$\frac{30}{12} = \frac{30}{12}$ ✓

Challenge
11-2 Use the Power of Algebra!

An equation may be used to solve a problem involving angle measure in a triangle.

In isosceles triangle ABC, the measure of vertex angle C is 30° more than the measure of each base angle. Find the measure of each angle of the triangle.

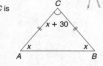

Let x = the number of degrees in m∠A.
Then x = the number of degrees in m∠B.
And $x + 30$ = the number of degrees in m∠C.

The sum of the measures of the angles of a triangle is 180°.

$x + x + x + 30 = 180$
$3x + 30 = 180$ Combine like terms.
$\underline{-30 \quad -30}$ Subtract 30.   Check:
$3x = 150$ m base ∠ = 50°
$\frac{3x}{3} = \frac{150}{3}$ Divide by 3.   m base ∠ = 50°
$x = 50$ ← m base ∠   m vertex ∠ = 80°
$x + 30 = 80$ ← m vertex ∠   180° ✓

So, the measure of each base angle is 50° and the measure of the vertex angle is 80°.

Write and solve an equation to find the measures of the angles of each triangle.

1. The measure of each of the base angles of an isosceles triangle is 9° less than 4 times the measure of the vertex angle.

$x + 4x - 9 + 4x - 9 = 180$
$9x - 18 = 180$
$\underline{+18 \quad +18}$
$9x = 198$
$\frac{9x}{9} = \frac{198}{9}$
$x = 22$
$4x - 9 = 79$

measure of each base angle = __79°__
measure of vertex angle = __22°__

2. The measure of the vertex angle of an isosceles triangle is one-fourth that of a base angle.

$x + x + \frac{x}{4} = 180$
$4 \cdot x + 4 \cdot x + 4 \cdot \frac{x}{4} = 4 \cdot 180$
$9x = 720$
$\frac{9x}{9} = \frac{720}{9}$
$x = 80$
$\frac{x}{4} = 20$

measure of each base angle = __80°__
measure of vertex angle = __20°__

Problem Solving
11-2 Solving Multi-Step Equations

A taxi company charges $2.25 for the first mile and then $0.20 per mile for each mile after the first, or $F = \$2.25 + \$0.20(m - 1)$ where F is the fare and m is the number of miles.

1. If Juan's taxi fare was $6.05, how many miles did he travel in the taxi?

   __20 miles__

2. If Juan's taxi fare was $7.65, how many miles did he travel in the taxi?

   __28 miles__

A new car loses 20% of its original value when you buy it and then 8% of its original value per year, or $D = 0.8V - 0.08Vy$ where D is the value after y years with an original value V.

3. If a vehicle that was valued at $20,000 new is now worth $9,600, how old is the car?

   __4 years__

4. A 6-year old vehicle is worth $12,000. What was the original value of the car?

   __$37,500__

The equation used to estimate typing speed is $S = \frac{1}{5}(w - 10e)$, where S is the accurate typing speed, w is the number of words typed in 5 minutes and e is the number of errors. Choose the letter of the best answer.

5. Jane can type 55 words per minute (wpm). In 5 minutes, she types 285 words. How many errors would you expect her to make?
   A 0   C 2
   Ⓑ 1   D 5

6. If Alex types 300 words in 5 minutes with 5 errors, what is his typing speed?
   F 48 wpm   H 59 wpm
   Ⓖ 50 wpm   J 60 wpm

7. Johanna receives a report that says her typing speed is 65 words per minute. She knows that she made 4 errors in the 5-minute test. How many words did she type in 5 minutes?
   A 285   Ⓒ 365
   B 329   D 1825

8. Cecil can type 35 words per minute. In 5 minutes, he types 255 words. How many errors would you expect her to make?
   F 2   H 6
   G 4   Ⓙ 8

Reading Strategies
11-2 Compare and Contrast

To solve equations with many terms, you can compare terms and combine the terms that are alike on one side of the = sign.

            term term term term
             ↓   ↓   ↓   ↓
            $3x + 7 + 4x - 3 = 32$

Step 1: Rearrange terms. $3x + 4x + 7 - 3 = 32$

Step 2: Combine like terms. $7x + 4 = 32$

Compare the first line of the equation with the second line.

1. How many terms are on the left of the = sign in the equation?

   __4 terms__

2. How many terms are there after the terms are rearranged in Step 1?

   __4 terms__

3. How are these two arrangements alike?

   __Possible answer: There are 4 terms in each equation.__

4. How are these two arrangements different?

   __Possible answer: The terms are in a different order.__

Now compare after the like terms are combined.

5. How many terms are on the left of the = sign in Step 2?

   __2 terms__

6. What is still alike and what is different about the equation now?

   __Sample answer: The equations have the same solution.__
   __The number of terms has changed.__

LESSON 11-2 Puzzles, Twisters & Teasers
Don't Miss the Boat!

Decide whether or not the given solution to each equation is correct. Circle the letter above your answer. Then unscramble the letters to solve the riddle.

1. $2x + 4 + 5x - 8 = 24$ $x = 4$
   - **A** correct
   - **Q** incorrect

2. $\frac{5n}{6} - \frac{1}{4} = \frac{3}{8}$ $n = 7$
   - **P** correct
   - **D** incorrect

3. $\frac{3y}{7} + \frac{5}{7} = -\frac{1}{7}$ $y = -2$
   - **M** correct
   - **J** incorrect

4. $\frac{x}{2} + \frac{2}{3} = \frac{5}{6}$ $x = 3$
   - **Y** correct
   - **I** incorrect

5. $5y - 2 - 8y = 31$ $y = 20$
   - **Z** correct
   - **R** incorrect

6. $\frac{2p}{3} + \frac{p}{4} - \frac{1}{6} = \frac{7}{2}$ $p = 4$
   - **A** correct
   - **X** incorrect

7. $\frac{b}{6} + \frac{3b}{8} = \frac{5}{12}$ $b = 9$
   - **T** correct
   - **L** incorrect

8. $2a + 7 + 3a = 32$ $a = 5$
   - **B** correct
   - **L** incorrect

9. $\frac{h}{6} + \frac{h}{8} = 1\frac{1}{6}$ $h = 2$
   - **W** correct
   - **Y** incorrect

10. $b + \frac{b}{5} - 10 = 4\frac{2}{5}$ $b = 12$
    - **R** correct
    - **B** incorrect

Who's the head of the seagull navy?

A D M I R A L   B Y R D

LESSON 11-3 Practice A
Solving Equations with Variables on Both Sides

Tell which term you would add or subtract on both sides of the equation so that the variable is only on one side.

1. $7x - 1 = 2x + 5$
   Subtract $2x$ from both sides.

2. $3y + 1 = 4y - 6$
   Subtract $3y$ from both sides.

3. $10 - y = 2y + 3$
   Add y to both sides.

4. $6x + 2 = 1 - x$
   Add x to both sides.

Solve.

5. $3a - 10 = 4 + a$ $a = 7$

6. $5y + 4 = 4y + 5$ $y = 1$

7. $b + 3 = 2b - 7$ $b = 10$

8. $2x - 1 = x - 3$ $x = -2$

9. $2w + 3 = 4w - 5$ $w = 4$

10. $6n + 4 = n - 11$ $n = -3$

11. $\frac{19m}{5} - \frac{3}{5} = 1.4 - \frac{m}{5}$ $m = 0.5$

12. $11 - \frac{t}{2} = \frac{t}{3} - 14$ $t = 30$

13. $29r + 4 = 28 + 13r$ $r = 1.5$

14. The difference of four times a number and five is the same as three more than twice a number. Find the number.   4

15. Fifteen more than twice the hours Carla worked last week is the same as three times the hours she worked this week decreased by 15. She worked the same number of hours each week. How many hours did she work each week?   30 hours

LESSON 11-3 Practice B
Solving Equations with Variables on Both Sides

Solve.

1. $7x - 11 = -19 + 3x$ $x = -2$

2. $11a + 9 = 4a + 30$ $a = 3$

3. $4t + 14 = \frac{6t}{5} + 7$ $t = -2.5$

4. $19c + 31 = 26c - 74$ $c = 15$

5. $\frac{3y}{8} - 9 = 13 + \frac{y}{8}$ $y = 88$

6. $\frac{3k}{5} + 44 = \frac{12k}{25} + 8$ $k = -300$

7. $10a - 37 = 6a + 51$ $a = 22$

8. $5w + 9.9 = 4.8 + 8w$ $w = 1.7$

9. $15 - x = 2(x + 3)$ $x = 3$

10. $15y + 14 = 2(5y + 6)$ $y = -0.4$

11. $14 - \frac{w}{8} = \frac{3w}{4} - 21$ $w = 40$

12. $\frac{1}{2}(6x - 4) = 4x - 9$ $x = 7$

13. $4(3d - 2) = 8d - 5$ $d = \frac{3}{4}$

14. $\frac{y}{3} + 11 = \frac{y}{2} - 3$ $y = 84$

15. $\frac{2x - 9}{3} = 8 - 3x$ $x = 3$

16. Forty-eight decreased by a number is the same as the difference of four times the number and seven. Find the number.   11

17. The square and the equilateral triangle at the right have the same perimeter. Find the length of the sides of the triangle.   12 units

LESSON 11-3 Practice C
Solving Equations with Variables on Both Sides

Solve.

1. $9x - 22 = 23 + 4x$ $x = 9$

2. $\frac{2y}{3} - 9 = 2y + 3$ $y = -9$

3. $5x - 36 = 24 + 2x$ $x = 20$

4. $\frac{7n}{9} - 62 = \frac{5n}{9} - 48$ $n = 63$

5. $\frac{x + 4}{5} = \frac{x - 6}{7}$ $x = -29$

6. $8(k + 6) = 3(k + 33)$ $k = 10.2$

7. $\frac{2x - 1}{3} = 4x + 3$ $x = -1$

8. $7d + 13 = 35 - 4d$ $d = 2$

9. $\frac{r}{5} - 26 = \frac{r}{2} - 29$ $r = 10$

10. $\frac{w}{4} + 5 = \frac{w}{3} + 10$ $w = -60$

11. $\frac{2m}{5} + 17 = \frac{m}{2} - 7$ $m = 240$

12. $5(5a + 3) = 14a - 29$ $a = -4$

13. $\frac{2x - 9}{3} = 10 - \frac{1}{5}x$ $x = 15$

14. $5m - 6 = 2.5m - 42$ $m = -14.4$

15. $\frac{3n - 7}{-2} = 5 - 0.25n$ $n = -1.2$

16. The sum of five and eight times a number is the same as fifty plus one-half the number. Find the number.   6

17. Jack and Jessica earned the same amount last week. They both work for the same hourly rate. Jack worked eighteen hours and had $42 deducted from his pay. Jessica worked fifteen hours and had $18 deducted from her pay. What was each person's salary last week after deductions? $102

LESSON 11-3 Reteach
Solving Equations with Variables on Both Sides

If there are variable terms on both sides of an equation, first collect them on one side. Do this by adding or subtracting.

If possible, collect the variable terms on the side where the coefficient will be positive.

$5x = 2x + 12$
$-2x \;\; -2x$ To collect on left side, subtract $2x$.
$3x = 12$
$\frac{3x}{3} = \frac{12}{3}$ Divide by 3.
$x = 4$

Check: Substitute into the original equation.
$5x = 2x + 12$
$5(4) \stackrel{?}{=} 2(4) + 12$
$20 \stackrel{?}{=} 8 + 12$
$20 = 20$ ✓

$-6z + 28 = 9z - 2$
$+6z \;\;\;\;\;\;\; +6z$ To collect on right side, add $6z$.
$28 = 15z - 2$
$+2 \;\;\;\;\;\; +2$ Add 2.
$30 = 15z$
$\frac{30}{15} = \frac{15z}{15}$ Divide by 15.
$2 = z$

Check: $-6z + 28 = 9z - 2$
$-6(2) + 28 \stackrel{?}{=} 9(2) - 2$
$-12 + 28 \stackrel{?}{=} 18 - 2$
$16 = 16$ ✓

Complete to solve and check each equation.

1. $9m = 4m - 25$
$-4m \;\; -4m$ To collect on left, subtract.
$5m = -25$
$\frac{5m}{5} = \frac{-25}{5}$ Divide.
$m = -5$

Check: $9m = 4m - 25$
$9(\underline{-5}) \stackrel{?}{=} 4(\underline{-5}) - 25$
$\underline{-45} \stackrel{?}{=} \underline{-20} - 25$
$\underline{-45} = \underline{-45}$ ✓

2. $3h - 7 = 5h + 1$
$-3h \;\;\;\;\;\; -3h$ To collect on right, subtract.
$-7 = \underline{2} h + 1$
$\underline{-1} \;\;\;\;\; \underline{-1}$ Subtract.
$\underline{-8} = \underline{2} h$
$\frac{-8}{2} = \frac{2h}{2}$ Divide.
$\underline{-4} = h$

Check: $3h - 7 = 5h + 1$
$3(\underline{-4}) - 7 \stackrel{?}{=} 5(\underline{-4}) + 1$
$\underline{-12} - 7 \stackrel{?}{=} \underline{-20} + 1$
$\underline{-19} = \underline{-19}$ ✓

LESSON 11-3 Reteach
Equations with Variables on Both Sides (continued)

To solve multi-step equations with variables on both sides: 1) *clear* fractions, 2) *combine* like terms, 3) *collect* variable terms on one side, and 4) *isolate* the variable by using properties of equality.

$\frac{t}{3} - \frac{5t}{6} + \frac{1}{2} = t - 1$ To clear fractions, determine LCD = 6.
$6 \cdot \frac{t}{3} - 6 \cdot \frac{5t}{6} + 6 \cdot \frac{1}{2} = 6 \cdot t - 6 \cdot 1$ Multiply *every* term by the LCD.
$2t - 5t + 3 = 6t - 6$ The fractions are cleared.
$-3t + 3 = 6t - 6$ Combine like terms.
$+3t \;\;\;\;\;\; +3t$ To collect variable terms, add $3t$ to both sides.
$3 = 9t - 6$
$+6 \;\;\;\;\; +6$ Add 6 to both sides.
$9 = 9t$
$\frac{9}{9} = \frac{9t}{9}$ Divide both sides by 9.
$1 = t$

Complete to solve.

3. $\frac{w}{4} + \frac{w}{2} + \frac{1}{4} = w$ Find the LCD of 2 and 4.
$4 \cdot \frac{w}{4} + 4 \cdot \frac{w}{2} + 4 \cdot \frac{1}{4} = \underline{4} \cdot w$ Multiply *every* term by the LCD.
$\underline{1} w + \underline{2} w + \underline{1} = \underline{4} w$
$\underline{3} w + \underline{1} = \underline{4} w$ Combine like terms.
$-3w \;\;\;\;\;\; -3w$ Subtract.
$\underline{1} = w$

4. $3m + 17 - m = 10 - m - 2$
$\underline{2} m + \underline{17} = \underline{8} - m$
$+m \;\;\;\;\;\;\; +m$
$\underline{3} m + 17 = \underline{8}$
$-17 \;\; -17$
$3m = \underline{-9}$
$m = \underline{-3}$

Check: $3m + 17 - m = 10 - m - 2$
$3(\underline{-3}) + 17 - (\underline{-3}) \stackrel{?}{=} 10 - (\underline{-3}) - 2$
$-9 + 17 + 3 \stackrel{?}{=} 10 + 3 - 2$
$\underline{8} + \underline{3} \stackrel{?}{=} \underline{13} - 2$
$11 = 11$ ✓

LESSON 11-3 Challenge
A Handy Tool!

A **lever** is a bar that can turn about a fixed point called the **fulcrum**.

The ancient Greek mathematician Archimedes knew the power of the *lever principle*. He has been quoted as saying "Give me a place to stand and I will move the Earth."

The Lever Principle

A weight w_1 is placed on one arm of a lever at a distance d_1 from the fulcrum. A second weight w_2 is placed on the other arm at a distance d_2 from the fulcrum.
$w_1 \cdot d_1 = w_2 \cdot d_2$

This equation may be used to solve a problem involving the lever principle.

A 14-foot plank is used as a lever with a 120-lb box on one end and a 90-lb box on the other end. If the boxes balance one another, how far from the fulcrum is each box?

Let x = 120-lb box's distance from fulcrum. Then $14 - x$ = 90-lb box's distance from fulcrum.

$w_1 \cdot d_1 = w_2 \cdot d_2$
$120 \cdot x = 90 \cdot (14 - x)$
$120x = 1260 - 90x$
$+90x \;\;\;\;\;\; +90x$
$210x = 1260$
$\frac{210x}{210} = \frac{1260}{210}$
$x = 6$ ft ← 120-lb box's distance from the fulcrum
$14 - x = 8$ ft ← 90-lb box's distance from the fulcrum

Write and solve an equation.

A 21-ft plank is used as a lever with a 108-lb barrel on one end and a 81-lb barrel on the other end. If the barrels balance one another, how far from the fulcrum is the 108-lb barrel?

The 108-lb barrel is __9 ft__ from the fulcrum.

Let x = 108-lb barrel's distance from fulcrum.
Then $21 - x$ = 81-lb barrel's distance.
$108x = 81(21 - x)$
$108x = 1701 - 81x$
$189x = 1701$
$x = 9$

LESSON 11-3 Problem Solving
Solving Equations with Variables on Both Sides

The chart below describes three long-distance calling plans. Round to the nearest minute. Write the correct answer.

Long-Distance Plans		
Plan	Monthly Access Fee	Charge per minute
A	$3.95	$0.08
B	$8.95	$0.06
C	$0	$0.10

1. For what number of minutes will plan A and plan B cost the same?
250 minutes

2. For what number of minutes per month will plan B and plan C cost the same?
224 minutes

3. For what number of minutes will plan A and plan C cost the same?
198 minutes

Choose the letter for the best answer.

4. Carpet Plus installs carpet for $100 plus $8 per square yard of carpet. Carpet World charges $75 for installation and $10 per square yard of carpet. Find the number of square yards of carpet for which the cost including carpet and installation is the same.
A 1.4 yd²
(C) 12.5 yd²
B 9.7 yd²
D 87.5 yd²

5. One shuttle service charges $10 for pickup and $0.10 per mile. The other shuttle service has no pickup fee but charges $0.35 per mile. Find the number of miles for which the cost of the shuttle services is the same.
F 2.5 miles
G 22 miles
(H) 40 miles
J 48 miles

6. Joshua can purchase tile at one store for $0.99 per tile, but he will have to rent a tile saw for $25. At another store he can buy tile for $1.50 per tile and borrow a tile saw for free. Find the number of tiles for which the cost is the same. Round to the nearest tile.
A 10 tiles
C 25 tiles
B 13 tiles
(D) 49 tiles

7. One plumber charges a fee of $75 per service call plus $15 per hour. Another plumber has no flat fee, but charges $25 per hour. Find the number of hours for which the cost of the two plumbers is the same.
F 2.1 hours
(H) 7.5 hours
G 7 hours
J 7.8 hours

LESSON 11-3 Reading Strategies: Follow a Procedure

Equations may have variables on both sides. Follow these steps to get the variables on one side of the equation.

Solve $6x - 7 = 2x + 5$.

Step 1: Get all variables on one side of the equation.
$6x - 2x - 7 = 2x - 2x + 5$ Subtract $2x$ from both sides.
$4x - 7 = 5$

Step 2: Get all constants on the other side of the equation.
$4x - 7 + 7 = 5 + 7$ Add 7 to both sides.
$4x = 12$

Step 3: Solve.
$\frac{4x}{4} = \frac{12}{4}$ Divide both sides by 4.
$x = 3$

Use the above procedure to answer each question.

1. What is the first step to solve equations with variables on both sides?
   Get the variables on one side of the equation.

2. What was done to get the variables on one side?
   $2x$ was subtracted from both sides.

3. Write the equation with the variables on one side only.
   $4x - 7 = 5$

4. What is the second step in solving the equation?
   Get all constants on the other side of the equation.

5. What was done to get the constants on one side?
   7 was added to both sides of the equation.

6. What was the last step to solve the equation?
   Both sides of the equation were divided by 4.

LESSON 11-3 Puzzles, Twisters & Teasers: Getting a New CD!

You've just earned some extra money and you want to buy a new CD, so you need to get to the music store. Start by solving the equations below. Once you have the solutions follow the directions to work your way through the maze.

1. $5x - 2 = x + 6$ $x = \underline{2}$ Start at the S and go right x spaces.
2. $6k - 6 = 6 + 4k$ $k = \underline{6}$ Go down k spaces.
3. $2a + 3 = 3a - 2$ $a = \underline{5}$ Go right a spaces.
4. $4(t - 5) + 2 = t + 3$ $t = \underline{7}$ Go right t spaces.
5. $2c + 4 - 3c = -9 + c + 5$ $c = \underline{4}$ Go up c spaces.
6. $5n - 3 = 2n + 12$ $n = \underline{5}$ Go right n spaces.
7. $3d + 4 = d + 18$ $d = \underline{7}$ Go down d spaces.

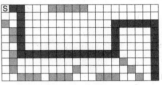

LESSON 11-4 Practice A: Solving Inequalities by Multiplying or Dividing

1. $4x > -20$
   $x > -5$

2. $3 \geq \frac{y}{5}$
   $y \leq 15$

3. $-\frac{b}{8} \geq 3$
   $b \leq -24$

4. $-6d < 18$
   $d > -3$

5. $63 \geq 7f$
   $f \leq 9$

6. $-\frac{g}{4} \leq 2$
   $g \geq -8$

7. $13 < \frac{h}{3}$
   $h > 39$

8. $-7j > -14$
   $j < 2$

9. Cheryl wants to buy a bicycle that costs $160. If she saves $12 each week, what is the fewest number of weeks she must save in order to buy the bicycle?
   at least 14 weeks

10. Mark worked on math homework less than $\frac{1}{3}$ the amount of time that his brother did. If Mark spent 25 minutes on his math homework, how much time did his brother spend on his math homework?
    more than 75 minutes, or 1 hour 15 minutes

LESSON 11-4 Practice B: Solving Inequalities by Multiplying or Dividing

Solve and graph.

1. $\frac{m}{-5} \leq 4$
   $m \geq -20$

2. $-16 < -8n$
   $n < 2$

3. $7p \geq 49$
   $p \geq 7$

4. $10 > \frac{q}{2}$
   $q < 20$

5. $-\frac{r}{3} \leq 15$
   $r \geq -45$

6. $22 > -2s$
   $s > -11$

7. $-6t < -24$
   $t > 4$

8. $\frac{v}{20} \geq 2$
   $v \geq 40$

9. On a snorkeling trip, Antonia dove at least 7 times as deep as Lucy did. If Antonia dove 35 feet below the ocean's surface, what was the deepest that Lucy dove?
   5 feet

10. Last week, Saul ran more than one-fifth the distance that his friend Omar ran. If Saul ran 14 miles last week, how far did Omar run?
    less than 70 miles

LESSON 11-4 Practice C
Solving Inequalities by Multiplying or Dividing

Solve and graph.

1. $18 \geq \frac{b}{-3}$

   $b \geq 54$

2. $6d > 42$

   $d > 7$

3. $5f < -15$

   $f < -3$

4. $24 \leq \frac{g}{2}$

   $g \geq 48$

5. $-4 < \frac{h}{-2}$

   $h < 8$

6. $8j \geq -40$

   $j \geq -5$

7. $\frac{k}{6} \geq 3$

   $k \geq 18$

8. $-56 < -7m$

   $m < 8$

9. Keisha told her swim coach that she would swim at least half a mile. If one lap is 50 yards, what is the fewest number of laps she must swim?

   18 laps

10. Mr. Wallace measured the length of a board that he was going to cut into 16 equal parts. If each part had to be less than 4 inches long, how long could the board be?

    less than 64 inches

LESSON 11-4 Reteach
Solving Inequalities by Multiplying or Dividing

To solve an inequality, multiply and divide the same way you would solve an equation. But, if you multiply or divide by a negative number, you must reverse the inequality sign.

Divide by a Positive Number

$2x < 14$

$\frac{2x}{2} < \frac{14}{2}$

$x < 7$

Divide by a Negative Number

$-2x < 14$

$\frac{-2x}{-2} > \frac{14}{-2}$ Reverse the inequality sign.

$x > -7$

To check your solution, choose two numbers from the graph and substitute them into the original equation. Choose a number that should be a solution and a number that should not be a solution.

Check

According to the graph, -6 should be a solution, but -8 should not be.

$-2x < 14$ $-2x < 14$

$\frac{-2 \cdot -8}{-2} \overset{?}{>} \frac{14}{-2}$ $\frac{-2 \cdot -6}{-2} \overset{?}{>} \frac{14}{-2}$

$-8 > -7$ ✗ $-6 > -7$ ✓

Complete to solve. Then graph the equation and check.

1. $-3y \geq 24$

   $\frac{-3y}{-3} \leq \frac{24}{-3}$

   $y \leq -8$

   Values used to check solution will vary, but should include one number ≤ -8 and one number > -8.

2. $\frac{s}{-9} < 4$

   $-9 \cdot \frac{s}{-9} > -9 \cdot 4$

   $s > -36$

   Values used to check solution will vary, but should include one number ≤ -36 and one number > -36.

LESSON 11-4 Challenge
Compounding the Problem

Inequalities that have more than one inequality sign are compound inequalities. You can solve them the same way you solve other inequalities. But, you must check to be sure the solution makes sense.

Example 1

$4 < 2x < 10$

$\frac{4}{2} < \frac{2x}{2} < \frac{10}{2}$ Divide each part by 2.

$2 < x < 5$ This means that $x > 2$ and $x < 5$. That makes sense.

Example 2

$-4 \geq \frac{x}{-3} \geq 2$

$-3 \cdot -4 \leq -3 \cdot \frac{x}{-3} \leq -3 \cdot 2$ Multiply each part by -3. Reverse the inequality symbols.

$12 \leq x \leq -6$ This means that $x \geq 12$ and $x \leq -6$. That does not make sense, so the inequality has no solution.

Solve and graph. If the inequality has no solution, write no solution.

1. $5 > \frac{x}{5} > 3$

   $25 > x > 15$

2. $14 \leq 2y < 18$

   $7 \leq y < 9$

3. $20 < -5z < 35$

   $-4 > z > -7$

4. $1 \leq \frac{b}{-2} < -2$

   $-2 \geq b > 4$; no solution

LESSON 11-4 Problem Solving
Solving Inequalities by Multiplying or Dividing

Write the correct answer

1. A bottle contains at least 4 times as much juice as a glass contains. The bottle contains 32 fluid ounces. Write an inequality that shows this relationship.

   $4x \leq 32$

2. Solve the inequality in Exercise 1. What is the greatest amount the glass could contain?

   $x \leq 8$; 8 fluid ounces

3. In the triple jump, Katrina jumped less than one-third the distance that Paula jumped. Katrina jumped 5 ft 6 in. Write an inequality that shows this relationship.

   $\frac{x}{3} > 66$

4. Solve the inequality in Exercise 3. How far could Paula could have jumped?

   $x > 198$; more than 198 in., or 16 ft 6 in.

Choose the letter for the best answer.

5. Melinda earned at least 3 times as much money this month as last month. She earned $567 this month. Which inequality shows this relationship?

   A $567 < x$
   B $567 < 3x$
   C $567 > 3x$
   D $567 \geq 3x$

6. The shallow end of a pool is less than one-quarter as deep as the deep end. The shallow end is 3 feet deep. Which inequality shows this relationship?

   F $4 > 3x$
   G $4x < 3$
   H $\frac{x}{4} > 3$
   J $\frac{x}{4} < 3$

7. Arthur worked in the garden more than half as long as his brother. Arthur worked 6 hours in the garden. How long did his brother work in the garden?

   A less than 3 hours
   B 3 hours
   C less than 12 hours
   D more than 12 hours

8. The distance from Bill's house to the library is no more than 5 times the distance from his house to the park. If Bill's house is 10 miles from the library, what is the greatest distance his house could be from the park?

   F 2 miles
   G more than 2 miles
   H 20 miles
   J less than 20 miles

LESSON 11-4 Reading Strategies
Understand Symbols

If you know the meanings of the inequality symbols, you can read and write inequalities as word sentences, and you can write word sentences as inequalities.

- < less than
- > greater than, or more than
- ≤ less than or equal to, or no more than
- ≥ greater than or equal to, or at least

Inequality	Word Sentence
$6 < x$	Six is less than x.
$y > 14$	y is greater than fourteen, or y is more than fourteen.
$5 \leq z$	Fifteen is less than or equal to z, or fifteen is no more than z.
$b \geq 5$	b is greater than or equal to five, or b is at least 5.

Many inequalities include multiplication or division.

Inequality	Word Sentence
$21 < 3x$	Twenty-one is less than three times x.
$\frac{y}{3} > 8$	y divided by three is greater than eight.
$4 \leq 2z$	Four is less than or equal to two times z, or four is no more than $2z$.
$\frac{b}{4} \geq 10$	b divided by four is greater than or equal to ten, or b divided by four is at least ten.

Write the inequality as a word sentence.

1. $5d > 40$
   Five times d is greater than 40, or five times d is more than 40.

2. $\frac{f}{6} \leq 3$ f divided by six is less than or equal to three, or f divided by six is no more than 3.

3. $11 < \frac{g}{2}$ Eleven is less than g divided by 2.

4. $16 \geq 4h$ Sixteen is greater than or equal to four times h, or sixteen is at least 4 times h.

Write an inequality that you could use to solve the problem.

5. A tree is more than five times as tall as a math student. The tree is 28 feet tall.
   Variables will vary. Possible answer: $28 > 5x$

LESSON 11-4 Puzzles, Twisters, and Teasers
Other Things Being Equal

Complete to solve each inequality. Use the key to match each answer with its letter. Unscramble the letters to answer the riddle.

Key

A	B	C	D	E	F	G	H	I	J	K	L	M
1	2	3	4	>	6	<	8	9	10	11	12	13
N	O	P	Q	R	S	T	U	V	W	X	Y	Z
14	15	16	≤	18	19	20	21	22	23	24	25	≥

1. $\frac{x}{-6} > -2$ → $x < \underline{\ 12\ }$

2. $\frac{j}{-3} < -7$ → $j > \underline{\ 21\ }$

3. $400 \leq 400y$ → $y \geq \underline{\ 1\ }$

4. $65 > -13d$ → $d \underline{\ >\ } -5$

5. $\frac{t}{7} < 2$ → $t < \underline{\ 14\ }$

6. $17g \leq -68$ → $g \underline{\ \leq\ } -4$

7. $3 \geq \frac{h}{7}$ → $h \leq \underline{\ 21\ }$

How are all numbers created?

U N E Q U A L

LESSON 11-5 Practice A
Solving Two-Step Inequalities

Write yes or no to tell whether the inequality symbol would be reversed in the solution. Do not solve.

1. $2x - 4 < 20$ no
2. $4 - 3y \geq 21$ yes
3. $6x + 17 > 3$ no
4. $-\frac{a}{5} - 4 \geq -2$ yes

Solve.

5. $2x - 17 \geq 29$ $x \geq 23$
6. $8 - \frac{k}{2} < -12$ $k > 40$
7. $23 - 3w < -34$ $w > 19$
8. $24 - 0.6x < 60$ $x > -60$
9. $10 \leq 5 - 2d$ $d \leq -2.5$
10. $\frac{x}{3} + 5 \leq 14$ $x \leq 27$
11. $\frac{2}{3} \geq \frac{y}{6} - \frac{1}{2}$ $y \leq 7$
12. $\frac{-a}{7} + \frac{1}{7} > \frac{1}{14}$ $a < \frac{1}{2}$

Solve and graph.

13. $2x - 1 < 3$ $x < 2$

14. $16 \geq 1 - 3a$ $a \geq -5$

15. $\frac{y}{2} - \frac{3}{4} \leq \frac{1}{2}$ $y \leq \frac{5}{2}$, or $2\frac{1}{2}$

16. $\frac{d}{3} + \frac{5}{12} > \frac{1}{4}$ $d > -\frac{1}{2}$

17. Mrs. Ocosta is paid a 5% commission on her sales each week. In addition, she receives a base salary of $375. What should the amount of her sales be for the week if she hopes to make at least $600 this week?
    $4500

LESSON 11-5 Practice B
Solving Two-Step Inequalities

Solve and graph.

1. $4x - 2 < 26$ $x < 7$

2. $6 - \frac{1}{5}y \leq 7$ $y \geq -5$

3. $2x + 27 \geq 15$ $x \geq -6$

4. $10x > 14x + 8$ $x < -2$

5. $7 - 4w \leq 19$ $w \geq -3$

6. $\frac{k}{5} + \frac{3}{20} < \frac{3}{10}$ $k < \frac{3}{4}$

7. $4.8 - 9.6x \leq 14.4$ $x \geq -1$

8. $\frac{2}{9} + \frac{y}{3} > \frac{1}{3}$ $y > \frac{1}{3}$

9. One-third of a number, decreased by thirty-six, is at most twenty-two. Find the number. $n \leq 174$

10. Jack wants to run at least 275 miles before the baseball season begins. He has already run 25 miles. He plans to run 2.5 miles each day. At this rate, what is the fewest number of days he will need to reach his goal?   100 days

LESSON 11-5 Practice C
Solving Two-Step Inequalities
Solve and graph.

1. $18 - 8x > 2$

    $x < 2$

2. $-\frac{5}{6} \leq -\frac{m}{12} - \frac{3}{4}$

    $m \geq -1$

3. $-2(w - 6) > 2$

    $w < 5$

4. $\frac{2(a-6)}{-9} \geq 0$

    $a \leq 6$

5. $9(3x - 7) < 18$

    $x < 3$

6. $\frac{d}{4} - 2 \geq -1$

    $d \geq 4$

7. $\frac{3(x+2)}{-4} \leq 5 - 4x$

    $x \leq 2$

8. $\frac{4n-3}{5} < \frac{3(n-1)}{4}$

    $n < -3$

9. Deidre goes out for lunch and realizes that she only has $10 with her. If she wants to leave a 20% tip and knows she will have to pay 5% tax on her lunch, what is the most expensive lunch she can order?

    $8

10. Mr. and Mrs. Schaefer are selling their house. Their real estate agent's fee is 7% of the selling price. The Schaefers want to make at least $175,000 from the sale. To the nearest dollar, what must be the selling price of the house?

    The house must sell for at least $188,172.

LESSON 11-5 Reteach
Solving Two-Step Inequalities

To solve an inequality, undo operations the same way you would with an equation. But, when multiplying or dividing by a negative number, reverse the inequality symbol.

$3x + 2 > 11$	To undo addition,	$-3x + 2 > 11$	To undo addition,
$-2\ \ -2$	subtract 2.	$-2\ \ -2$	subtract 2.
$3x > 9$	To undo multiplication,	$-3x > 9$	To undo multiplication,
$\frac{3x}{3} > \frac{9}{3}$	divide by 3.	$\frac{-3x}{-3} < \frac{9}{-3}$	divide by -3 and
$x > 3$		$x < -3$	change $>$ to $<$.

The solution set contains all real numbers greater than 3.

The solution set contains all real numbers less than -3.

Complete to solve and graph.

1. $2t + 1 \leq 9$ To undo addition,

    $\underline{-1}\ \underline{-1}$ subtract.

    $2t \leq 8$ To undo multiplication,

    $\frac{2t}{2} \leq \frac{8}{2}$ divide.

    $t \leq 4$

2. $-2t + 1 \leq 9$ To undo addition,

    $\underline{-1}\ \underline{-1}$ subtract.

    $-2t \leq 8$ To undo multiplication,

    $\frac{-2t}{-2} \geq \frac{8}{-2}$ divide by -2 and

    $t \geq -4$ change $\leq$ to $\geq$.

    give credit for either $\leq$ or $\geq$

3. $-3z - 2 > 1$

    $\underline{+2}\ \underline{+2}$

    $-3z > 3$

    $\frac{-3z}{-3} < \frac{3}{-3}$

    $z < -1$

4. $3z - 2 > 1$

    $\underline{+2}\ \underline{+2}$

    $3z > 3$

    $\frac{3z}{3} > \frac{3}{3}$

    $z > 1$

LESSON 11-5 Reteach
Solving Two-Step Inequalities (continued)

To solve multistep inequalities, you may need to clear fractions. Multiply both sides by the LCD.

$\frac{v}{4} + \frac{1}{4} > -\frac{1}{2}$ The LCD is 4.

$4 \cdot \frac{v}{4} + 4 \cdot \frac{1}{4} > 4 \cdot -\frac{1}{2}$ Multiply by the LCD.

$v + 1 > -2$

$\underline{-1}\ \underline{-1}$ Subtract from both sides.

$v > -3$

Complete to solve and graph.

5. $-\frac{b}{4} - \frac{7}{12} \leq \frac{2}{3}$ Find the LCD.

    $12 \cdot -\frac{b}{4} - 12 \cdot \frac{7}{12} \leq 12 \cdot \frac{2}{3}$ Multiply by the LCD.

    $-3b - 7 \leq 8$

    $\underline{+7}\ \underline{+7}$ Add.

    $-3b \leq 15$

    $\frac{-3b}{-3} \geq \frac{15}{-3}$ Divide and change symbol.

    $b \geq -5$ Check direction.

6. $\frac{2}{7} + \frac{y}{14} \geq -\frac{1}{2}$

    $14 \cdot \frac{2}{7} + 14 \cdot \frac{y}{14} \geq 14 \cdot -\frac{1}{2}$

    $4 + y \geq -7$

    $-4\ \ \ \ -4$

    $y \geq -11$

7. $-\frac{1}{3} > \frac{x}{9} + \frac{2}{3}$

    $9 \cdot -\frac{1}{3} > 9 \cdot \frac{x}{9} + 9 \cdot \frac{2}{3}$

    $-3 > x + 6$

    $-6\ \ \ \ -6$

    $-9 > x$

    $x < -9$

LESSON 11-5 Challenge
Updated Pony Express

Pat wants to send some copies of her newly published book to friends. According to the U.S. Postal Service:

Rates are based on the weight of the piece and the zone (distance from origin to destination ZIP code).

The combined length and girth (perimeter of an end) of a package may not exceed 108 inches.

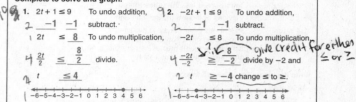

1. Pat wants the box that contains books to be 6 inches high, and twice as long as it is wide.

    Let x represent the width of a box that Pat might use.

    Write and solve an inequality to find all possible widths for a box that will satisfy the postal requirements and Pat's conditions.

    $2(x) + 2(6) + 2x \leq 108$
    $4x + 12 \leq 108$
    $4x \leq 96$
    $x \leq 24$

    possible width: ≤ 24 inches

2. Pat's husband, Mike, suggests that the box be 8 inches high and that the length be 3 times the width.

    Let z represent the length of a box that Mike suggests.

    Write and solve an inequality to find, to the nearest inch, the maximum length for a box that will satisfy.

    $2\left(\frac{z}{3}\right) + 2(8) + z \leq 108$
    $\frac{2z}{3} + 48 + 3z \leq 324$
    $2z + 48 + 3z \leq 324$
    $5z + 48 \leq 324$
    $5z \leq 276$
    $z \leq 55.2$

    maximum length: 55 inches

3. On May 1, 2002, Pat shipped a box containing a book to a friend who lives in Zone 4. Pat paid $2.08 to ship this package.

    According to the table below, write an inequality to show the weight of this package.

    $2.5 < x \leq 3$

Bound Printed Matter Rates

Weight Not Over (pounds)	Local, Zones 1&2	Zone 3	Zone 4	Zone 5	Zone 6	Zone 7	Zone 8
1.0	$1.80	$1.83	$1.87	$1.93	$1.99	$2.06	$2.21
1.5	1.80	1.83	1.87	1.93	1.99	2.06	2.21
2.0	1.84	1.88	1.94	2.02	2.10	2.19	2.38
2.5	1.90	1.95	2.00	2.11	2.21	2.33	2.57
3.0	1.94	2.00	2.08	2.20	2.32	2.46	2.75
3.5	1.99	2.06	2.15	2.29	2.43	2.60	2.93
4.0							

LESSON 11-5 Problem Solving
Solving Two-Step Inequalities

A school club is selling printed T-shirts to raise $650 for a trip. The table shows the profit they will make on each shirt after they pay the cost of production.

1. Suppose the club already has $150, at least how many 50/50 shirts must they sell to make enough money for the trip?

   91 shirts

Shirt	Profit
50/50	$5.50
100% cotton	$7.82

2. Suppose the club already has $100, but it plans to spend $50 on advertising. At least how many 100% cotton shirts must they sell to make enough money for the trip?

   77 shirts

3. Suppose the club sold thirty 50/50 shirts on the first day of sales. At least how many more 50/50 shirts must they sell to make enough money for the trip?

   89 shirts

For Exercises 4–5, use this equation to estimate typing speed, $S = \frac{w}{5} - 2e$, where S is the accurate typing speed, w is the number of words typed in 5 minutes, and e is the number of errors. Choose the letter for the best answer.

4. One of the qualifications for a job is a typing speed of at least 65 words per minute. If Jordan knows that she will be able to type 350 words in five minutes, what is the maximum number of errors she can make?
   A 0   C 3
   B 2   D 4

5. Tanner usually makes 3 errors every 5 minutes when he is typing. If his goal is an accurate typing speed of at least 55 words per minute, how many words does he have to be able to type in 5 minutes?
   F 61 words   **H 305 words**
   G 300 words   J 325 words

6. A taxi charges $2.05 per ride and $0.20 for each mile, which can be written as $F = \$2.05 + \$0.20m$. How many miles can you travel in the cab and have the fare be less than $10?
   A 15   **C 39**
   B 25   D 43

7. Celia's long distance company charges $5.95 per month plus $0.06 per minute. If Celia has budgeted $30 for long distance, what is the maximum number of minutes she can call long distance per month?
   F 375 minutes   H 405 minutes
   G 400 minutes   J 420 minutes

LESSON 11-5 Reading Strategies
Follow a Procedure

You can use these steps to help you solve a two-step inequality.

Solve $-8 < 4x + 4$.

Step 1: Get the variable by itself on one side of the inequality.
$-8 - 4 < 4x + 4 - 4$ Subtract 4 from both sides.
$-12 < 4x$

Step 2: Solve.
$-\frac{12}{4} < \frac{4x}{4}$ Divide both sides by 4.
$-3 < x$

Step 3: Rewrite the solution so the variable comes first.
$x > -3$

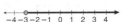

Use the procedure to answer each question.

1. What did the procedure tell you to do first?
   Get the variable by itself on one side of the inequality.

2. How did you get the variable by itself in this problem?
   Subtracted 4 from both sides

3. What is the second step given?
   Solve the inequality.

4. How did you solve this inequality?
   Divide both sides by 4

5. How would the graph for $x \geq -3$ be different than the above graph?
   Possible answer: The circle at −3 would be solid.

LESSON 11-5 Puzzles, Twisters & Teasers
Get to the Beach!

Decide whether or not the given solution to each inequality is correct. Circle the letter above your answer. Then unscramble the letters to solve the riddle.

1. $2x - 3 > 5$; $x > 4$
   (S) correct   A incorrect

   $3 > 3x - 6$; $3 > x$
   (I) correct   R incorrect

2. $-10 < 3x + 2$; $-4 < x$
   (A) correct   U incorrect

7. $10x > 200 + 2x$; $x > 25$
   (C) correct   V incorrect

3. $-2x + 4 \leq 3$; $x \geq \frac{1}{2}$
   N correct   W incorrect [H correct]

8. $3k - 2 > 13$; $k > 3$
   T correct   **(H)** incorrect

4. $-2 - x > 5$; $x < -7$
   (D) correct   T incorrect

9. $10x + 2 > 42$; $x < 4$
   I correct   **(E)** incorrect

5. $\frac{2x}{3} + \frac{1}{2} \leq \frac{5}{6}$; $x \leq \frac{1}{2}$
   (W) correct   A incorrect

10. $5 < -p - 12$; $p \geq 10$
    C correct   **(S)** incorrect

What do witches like to eat at the beach?

S A N D W I C H E S

LESSON 11-6 Practice A
Systems of Equations

Substitute to determine if the ordered pair (3, 2) is a solution of the following systems of equations. Write *yes* or *no*.

1. $x + y = 5$
   $2x - y = 4$
   yes

2. $y = x - 1$
   $y = 3x - 7$
   yes

3. $y = -x + 5$
   $x - 2y = -4$
   no

4. $4x - y = 10$
   $3y = x + 1$
   no

Solve each system of equations.

5. $y = x$
   $y = -x$
   (0, 0)

6. $y = 3x + 1$
   $y = 2x - 1$
   (−2, −5)

7. $y = 2x - 3$
   $y = x - 2$
   (1, −1)

8. $y = -2x + 3$
   $y = 3x + 3$
   (0, 3)

9. $x + y = 2$
   $2x + y = 1$
   (−1, 3)

10. $2x = y$
    $4x - y = -2$
    (−1, −2)

11. $2x + 4y = 24$
    $x - 2y = -8$
    (2, 5)

12. $x + y = 4$
    $6x + 2y = 8$
    There are an infinite number of solutions.

13. The sum of two numbers is 8. The difference of the two numbers is 2. Write a system of equations to find these two numbers and solve.
    $x + y = 8; x - y = 2; (5, 3)$

Practice B
11-6 Systems of Equations

Solve each system of equations.

1. $y = 2x - 4$
   $y = x - 1$
   __(3, 2)__

2. $y = -x + 10$
   $y = x + 2$
   __(4, 6)__

3. $y = 2x - 1$
   $y = -3x - 6$
   __(-1, -3)__

4. $y = 2x$
   $y = 12 - x$
   __(4, 8)__

5. $y = 2x - 3$
   $y = 2x + 1$
   __no solution__

6. $y = 3x - 1$
   $y = x + 1$
   __(1, 2)__

7. $x + y = 0$
   $5x + 2y = -3$
   __(-1, 1)__

8. $2x - 3y = 0$
   $2x + y = 8$
   __(3, 2)__

9. $2x + 3y = 6$
   $4x + 6y = 12$
   __infinite number__

10. $6x - y = -14$
    $2x - 3y = 6$
    __(-3, -4)__

11. The sum of two numbers is 24. The second number is 6 less than the first. Write a system of equations and solve it find the number.

    $x + y = 24$; $y = x - 6$; (15, 9)

15. Kerry and Luke biked a total of 18 miles in one weekend. Kerry biked 4 miles more than Luke. Write a system of equations and solve it to find how far each boy biked.

    $x + y = 18$; $x + 4 = y$; (7, 11); Luke biked 7 miles, and Kerry biked 11 miles

Practice C
11-6 Systems of Equations

Solve each system of equations.

1. $y = 3x - 1$
   $y = -2x$
   __$\left(\frac{1}{5}, -\frac{2}{5}\right)$__

2. $2x + 3y = 12$
   $x = 4y - 5$
   __(3, 2)__

3. $-x + 2y = 8$
   $4x + y = -5$
   __(-2, 3)__

4. $2x - y = 7$
   $3x - 4y = 8$
   __(4, 1)__

5. $y = -0.4x$
   $y = -0.8x + 2$
   __(5, -2)__

6. $2x - y = 4$
   $\frac{1}{2}y = x - 2$
   __There are an infinite number of solutions.__

7. $2y = 5x + 5$
   $7y - 3x = 32$
   __(1, 5)__

8. $y = \frac{1}{4}x - 1$
   $x + 4y = -4$
   __(0, -1)__

9. $2x + 3y = 3$
   $3y = 5 - 2x$
   __no solution__

10. $4x - y = 3$
    $4x + y = 1$
    __$\left(\frac{1}{2}, -1\right)$__

11. The sum of two numbers is 381. Their difference is 155. Find the two numbers. __(113, 268)__

12. The theater group sold adult and student tickets for the play. They sold a total of 560 tickets. Each adult ticket was $8 and each student ticket was $3.50. The group took in a total of $3166. How many of each type of ticket were sold for the play?

    __292 student tickets; 268 adult tickets__

13. The cost of 20 blank recordable CDs and 6 music CDs is $80.70. The cost of 30 blank recordable CDs and 4 music CDs is $66.30. Find the cost of each blank recordable CD and each music CD.

    __blank recordable CD: $0.75; music CD: $10.95__

Reteach
11-6 Systems of Equations

Two or more equations considered together form a **system of equations**. To solve a system of equations, you can use a method of substitution.

Solve the system: $y = 3x$
$y - 5x = 20$

Use the first equation to substitute for y in the second equation.	$y - 5x = 20$ $3x - 5x = 20$	second equation Replace y with $3x$.
Solve the resulting equation for x.	$-2x = 20$ $\frac{-2x}{-2} = \frac{20}{-2}$ $x = -10$	Combine like terms. Divide by -2.
Substitute the x-value into the first equation to get the corresponding y-value.	$y = 3x$ $y = 3(-10) = -30$	

Check: Substitute both values in each of the original equations.

$y = 3x$ $y - 5x = 20$ $x = -10$ and $y = -30$
$-30 \stackrel{?}{=} 3(-10)$ $-30 - 5(-10) \stackrel{?}{=} 20$ So, the solution of the
$-30 = -30$ ✓ $-30 + 50 \stackrel{?}{=} 20$ system is $(-10, -30)$.
   $20 = 20$ ✓

Solve and check this system.

1. $y = 2x$
   $6x + y = 16$

   Use the first equation to substitute for y in the second equation.
   $6x + \underline{2x} = 16$
   Solve the resulting equation for x.
   $\underline{8x = 16}$
   $\underline{\frac{8x}{8} = \frac{16}{8}}$
   $\underline{x = 2}$
   Substitute the x-value to get the corresponding y-value.
   $y = 2x$
   $y = 2(\underline{2}) = \underline{4}$

   Check both values in each of the original equations.
   $y = 2x$ $6x + y = 16$
   $\underline{4} \stackrel{?}{=} 2(\underline{2})$ $6(\underline{2}) + \underline{4} \stackrel{?}{=} 16$
   $4 = 4$ ✓ $12 + 4 \stackrel{?}{=} 16$
      $16 = 16$ ✓

   So, the ordered pair __(2, 4)__ is the solution of the system.

Reteach
11-6 Systems of Equations (continued)

Sometimes, you first have to solve one equation for a variable.
Solve the system: $y + 3x = 7$
$x + 2y = 4$

Solve the first equation for y.	$y + 3x = 7$ $\underline{-3x} \underline{-3x}$ $y = 7 - 3x$	Subtract $3x$.
Substitute for y in the second equation. Solve for x.	$x + 2y = 4$ $x + 2(7 - 3x) = 4$ $x + 14 - 6x = 4$ $-5x + 14 = 4$ $ -14 -14$ $\frac{-5x}{-5} = \frac{-10}{-5}$ $x = 2$	second equation Replace y with $7 - 3x$. Distributive property Combine like terms. Subtract 14. Divide by -5.
Substitute the x-value into the first equation to get the corresponding y-value.	$y + 3x = 7$ $y + 3(2) = 7$ $y + 6 = 7$ $y + 6 - 6 = 7 - 6$ $y = 1$	

Check: Substitute both values in each of the original equations.
$y + 3x = 7$ $x + 2y = 4$ $x = 2$ and $y = 1$
$1 + 3(2) \stackrel{?}{=} 7$ $2 + 2(1) \stackrel{?}{=} 4$ The solution of the
$7 = 7$ ✓ $4 = 4$ ✓   system is (2, 1).

Solve and check this system.

2. $y - 2x = 0$
   $x - 2y = 6$

   Solve the first equation for y.
   $y - 2x = 0$
   $y = \underline{2x}$
   Use the result to substitute for y in the second equation.
   $x - 2y = 6$
   $x - 2(\underline{2x}) = 6$
   Solve the resulting equation for x.
   $x = \underline{-2}$

   Substitute the x-value to get the corresponding y-value.
   $y = 2x$
   $y = 2(-2)$
   $y = \underline{-4}$
   Check:
   $y - 2x = 0$ $x - 2y = 6$

   So, the ordered pair __(-2, -4)__ is the solution of the system.

LESSON 11-6 Challenge: Different Strokes!

The first objective in solving a system of two equations in two variables is to get to one equation in one variable. You have seen how to do this by using substitution.

Now you will see how to solve the system at the right by using addition.

$$x + y = 7$$
$$2x - y = 2$$

Add the equations to eliminate y.
Solve for x.

$$x + y = 7$$
$$2x - y = 2$$
$$3x = 9$$
$$\frac{3x}{3} = \frac{9}{3}$$
$$x = 3$$

Return to any equation with x and y.
Substitute the x-value.
Solve for y.

$$x + y = 7$$
$$3 + y = 7$$
$$-3 \quad -3$$
$$y = 4$$

You should check the values by substituting into the original equations.

The solution for this system is (3, 4).

If the coefficients of one variable in a set of equations are equal but not opposite, you can use subtraction to eliminate that variable.

Use the space provided to solve each system using addition or subtraction to eliminate a variable. Check your results.

1. $2x + 3y = 7$
   $x - 3y = 8$
   $3x = 15$
   $\frac{3x}{3} = \frac{15}{3}$
   $x = 5$

   $x - 3y = 8$
   $5 - 3y = 8$
   $-5 \quad -5$
   $-3y = 3$
   $\frac{-3y}{-3} = \frac{3}{-3}$
   $y = -1$

   Solution: (5, −1)

2. $x + 3y = 7$
   $-x + 2y = 8$
   $5y = 15$
   $\frac{5y}{5} = \frac{15}{5}$
   $y = 3$

   $x + 3y = 7$
   $x + 3(3) = 7$
   $x + 9 = 7$
   $-9 \quad -9$
   $x = -2$

   Solution: (−2, 3)

3. $x - y = 9$
   $x + 4y = 24$
   $-5y = -15$
   $\frac{-5y}{-5} = \frac{-15}{-5}$
   $y = 3$

   $x - y = 9$
   $x - 3 = 9$
   $+3 \quad +3$
   $x = 12$

   Solution: (12, 3)

LESSON 11-6 Problem Solving: Systems of Equations

After college, Julia is offered two different jobs. The table summarizes the pay offered with each job. Write the correct answer.

Job	Yearly Salary	Yearly Increase
A	$20,000	$2500
B	$25,000	$2000

1. Write an equation that shows the pay y of Job A after x years.

   $y = 20,000 + 2500x$

2. Write an equation that shows the pay y of Job B after x years.

   $y = 25,000 + 2000x$

3. Is (8, 35,000) a solution to the system of equations in Exercises 1 and 2?

   no

4. Solve the system of equations in Exercises 1 and 2.

   (10, 45,000)

5. If Julia plans to stay at this job only a few years and pay is the only consideration, which job should she choose?

   Job B

A travel agency is offering two Orlando trip plans that include hotel accommodations and pairs of tickets to theme parks. Use the table below. Choose the letter for the best answer.

Trip	Number of nights	Pairs of theme park tickets	Cost
A	3	2	$415
B	5	4	$725

6. Find an equation about trip A where x represents the hotel cost per night and y represents the cost per pair of theme park tickets.

   A $5x + 2y = 415$ C $8x + 6y = 415$
   B $2x + 3y = 415$ **D** $3x + 2y = 415$

7. Find an equation about trip B where x represents the hotel cost per night and y represents the cost per pair of theme park tickets.

   F $5x + 4y = 725$
   G $4x + 5y = 725$
   H $8x + 6y = 725$
   J $3x + 4y = 725$

8. Solve the system of equations to find the nightly hotel cost and the cost for each pair of theme park tickets.

   A ($50, $105)
   B ($125 $20)
   C ($105, $50)
   D ($115, $35)

LESSON 11-6 Reading Strategies: Focus on Vocabulary

A system of equations is a set of equations with a common solution. Think of the solar system. All of the planets have the Sun in common.

Equation A: $y = 5x - 3$ ← Both equations have the
Equation B: $y = 3x + 1$ ← same variables, x and y.

The **common solution** of a system of equations that contains two variables is an ordered pair (x, y) that solves both equations.

(x, y)
$\downarrow \quad \downarrow$
$(2, 7)$ is a solution for $y = 5x - 3$.

Use the ordered pair (2, 7) to answer the following questions.

1. What is a system of equations?

   a set of equations with a solution in common

2. What are the variables in the set of equations above?

   x and y

3. Which number in the solution stands for x?

   2

4. Which number in the solution stands for y?

   7

5. Rewrite Equation A by substituting 2 and 7 for x and y.

   $7 = 5(2) - 3$

6. Why is (2, 7) a solution of $y = 5x - 3$?

   because 7 is equal to 10 − 3

7. Rewrite Equation B by substituting 2 and 7 for x and y.

   $7 = 3(2) + 1$

8. Is (2, 7) a solution for Equation B? Why or why not?

   Yes; 7 is equal to 6 + 1.

9. Is (2, 7) a solution for this system of equations? Why or why not?

   Yes; it is a solution for both equations.

LESSON 11-6 Puzzles, Twisters & Teasers: System Solution!

Solve the crossword puzzle.

Across

1. A system of ___ is a set of two or more equations that contain two or more variables.
3. Solutions to a system of equations can be written as ___ pairs.
5. When solving systems of equations, you should find values for all the ___.
7. To ___ is to replace a variable with a value.
8. To find one variable, substitute a ___ for the other variable.

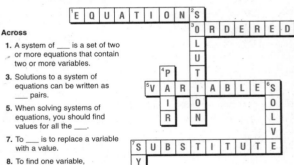

Down

2. A ___ of a system of equations is a set of values that are solutions of all the equations.
4. An ordered ___ may or may not be a solution of a system of equations.
6. It is easiest to ___ for a variable that has a coefficient of one.
7. To solve a general ___ of equations with two variables, solve both equations for one of the variables.

Crossword answers: 1-EQUATIONS, 2-SOLUTION, 3-ORDERED, 4-PAIR, 5-VARIABLES, 6-SOLVE, 7-SUBSTITUTE/SYSTEM, 8-VALUE